IEEE Recommended Practice for Industrial and Commercial Power Systems Analysis

Published by
The Institute of Electrical and Electronics Engineers, Inc

IEEE Recommended Practice for Industrial and Commercial Power Systems Analysis

Sponsor

Power System Technologies Committee
of the
IEEE Industry Applications Society

Approved December 20, 1979
IEEE Standards Board

Approved June 28, 1982
American National Standards Institute

IEEE Power Engineering Seminars

The IEEE sponsors seminars on the Color Books and other power engineering standards throughout the year.
Our seminars include:

- Protection and Co-Generation Plants Paralleled with Utility Transmission Systems
- Health Care Facilities Power Systems
- Planning, Design, Protection, Maintenance, and Operation of Industrial and Commercial Power Systems
- Electric Power Supply Systems for Nuclear Power Generating Stations
- Large Storage Batteries — Nickel-Cadmium and Lead

IEEE-sponsored seminars and training programs may also be brought to your plant. For details, write to the IEEE Standards Seminar Manager, 445 Hoes Lane, PO Box 1331, Piscataway, NJ 08855-1331 USA. In the US and Canada call us Toll Free at 1-800-678-IEEE and ask for Standards Seminars and Training Programs. Our fax number is 201-562-1571.

ISBN 0-471-09262-2

Sixth Printing
June 1990

Library of Congress Catalog Number 80-83820

© Copyright 1980 by

The Institute of Electrical and Electronics Engineers, Inc

November 24, 1980

SH07906

Foreword

(This Foreword is not a part of IEEE Std 399-1980, IEEE Recommended Practice for Industrial and Commercial Power Systems Analysis.)

This Recommended Practice is the product of about ten years of effort by a working group of the Power System Technologies Committee of IEEE Industry Applications Society. It is intended as a practical, general treatise on the theoretical basis of power system analysis, and as a reference work on the analytical techniques most commonly applied to electric power systems in industrial plants and commercial buildings.

IEEE Recommended Practice for Industrial And Commercial Power System Analysis, the IEEE Brown Book, joins the series of *Color Books* sponsored by the Power System Technologies Committee of IEEE Industry Applications Society. It is both complementary and supplementary to the other color books, extending the coverage of some topics which they introduce as well as discussing some entirely new material.

Comments, corrections, and suggestions for the next revision of the Brown Book are welcome and should be submitted to the

IEEE Standards Board
345 East 47th Street
New York, New York 10017.

At the time it recommended these practices, the working group of the Power System Technologies Committee had the following members and contributors:

Richard H. McFadden, *Chairman*

R. Gene Baggs
Graydon M. Bauer
Kao Chen
Lonnie E. Crawford
Leon C. Glahn
M. Shan Griffith
William R. Haack
John Lockhart
Russell O. Ohlson

A. Dewitt Patton
H. Paul Rouleau
David D. Shipp
David H. Smith
James F. Smith
J. R. Smith
George A. Terry
George W. Walsh
A. Jack Williams, Jr

Myron Zucker

Industrial and Commercial Power Systems Analysis

1st Edition

Working Group Members and Contributors

Richard H. McFadden, *Working Group Chairman*

Chapter 1 — **Introduction:** Kao Chen
Chapter 2 — **Applications of Power System Analysis:** Leon Glahn
Chapter 3 — **Analytical Procedures:** M. Shan Griffith
Chapter 4 — **System Modeling:** Paul Rouleau
Chapter 5 — **Load Flow Studies:** J. R. Smith
Chapter 6 — **Short Circuit Studies:** David H. Smith
Chapter 7 — **Stability Studies:** Richard H. McFadden
Chapter 8 — **Motor Starting Studies:** A. J. Williams, Jr and M. Shan Griffith
Chapter 9 — **Harmonic Studies:** David Shipp
Chapter 10 — **Switching Transients Studies:** George Walsh
Chapter 11 — **Reliability Studies:** A. D. Patton
Chapter 12 — **Grounding Mat Studies:** L. E. Crawford and M. Shan Griffith
Chapter 13 — **Computer Services:** John Lockhart

Contents

1. Introduction

1.1 General Discussion. IEEE Std 399-1980, IEEE Recommended Practice for Industrial and Commerical Power System Analysis, commonly known as the IEEE Brown Book, is published by the Institute of Electrical and Electronics Engineers (IEEE) as a reference source for plant electrical engineers in making power system studies. The IEEE Brown Book can also be helpful in preparing system modeling and data acquisition for an outside engineering consultant to perform necessary engineering studies prior to designing a new system or expanding an existing power system. Such information will help ensure high standards of power system reliability and maximize the utilization of capital investment.

The IEEE Brown Book has been prepared on a voluntary basis by engineers and designers functioning as the Power System Analysis Working Group within the IEEE, under the Industrial Power Systems Department of The Industry Applications Society. This recommended practice is not intended as a replacement for the many excellent texts available in this field. The IEEE Brown Book comple-ments the other IEEE color book standards, and emphasizes up-to-date techniques in system studies which are most applicable to industrial and commercial power systems. Today, such techniques are mostly computer oriented.

1.2 History of Power System Studies. The planning, design, and operation of a power system require continual and comprehensive analyses to evaluate current system performance and to ascertain the effectiveness of alternative plans for system expansion.

The computational work to determine power flows and voltage levels resulting from a single operating condition for even a small network is all but insurmountable if performed by manual methods. The need for computational aids led to the design of a special purpose analog computer (AC Network Analyzer) as early as 1929. It provided the ability to determine load flows and system voltage during normal and emergency conditions and to study the transient behavior of the system resulting from fault conditions and switching operations.

The earliest application of digital computers to power system problems dates back to the late 1940s. Most of the early applications were limited in scope, however, because of the small capacity of the punched card calculators in use during that period. The large scale digital computers became available in the middle 1950s. The initial success of the load flow program led to the development of programs for short circuit and transient stability calculations.

Today, the digital computer is an indispensable tool in power system planning where it is necessary to predict future growth and simulate day-to-day operations over periods of 20 years or more.

1.3 Applying Power System Analysis Techniques to Industrial and Commercial Power Systems.

As computer technology has advanced, so has the complexity of industrial and commercial power systems. These power systems have grown in recent decades with capacities far exceeding that of a small electric utility system.

Today, plant or building management personnel are highly concerned with system interruptions and their effect on overall operation. Therefore, they demand assurances of maximum return on capital investment for any sizable expansion. The complexity of modern industrial and commercial power systems has made manual performance of power system studies difficult and time consuming, if not impossible. However, through the use of digital computers these studies can be made with relative ease. Answers to many perplexing questions regarding impact of expansion on the system, short circuit capacity, stability, load distribution, etc, can be intelligently obtained.

The recent advances in computer technology now enable engineers to communicate with the computer and change or modify the system to meet whatever design criteria may arise.

1.4 Purposes of this Recommended Practice

1.4.1 Why a Study. As is stated in Section 2, the planning, design, and operation of industrial or commercial power systems require several studies to assist in the evaluation of the initial and future system performance, reliability, safety, and ability to grow with production or operating requirements. The studies most likely needed are load-flow studies, short-circuit studies, stability studies, and motor-starting studies. A brief summary of such studies and what each can accomplish are given in Section 2 to help a plant engineer determine overall system needs before he proceeds with specific planning and engineering design.

Additional studies of transients, reliability, grounding, and harmonics may also be required. The plant engineer in charge of system design must decide which studies are needed to ensure that the system will operate safely, economically, and efficiently over the expected life of the system.

1.4.2 How to Prepare for a Power System Study. For a plant engineer to solve a power system analysis problem, he must be thoroughly familiar with valid electrical engineering fundamentals, such as the Thevenin equivalent circuit, the phasor representation, the Fourier series, symmetrical components, etc. He can then analyze the problem and prepare necessary equivalent circuits and basic system data before proceeding with a computer program. Failure to use a valid analytical procedure to establish a sound basic approach to the problem could lead to disastrous consequences in both the design and operation of a system.

Section 3 offers an excellent review of the most essential fundamentals in a system study.

To set up a computer program for system analysis, certain basic data must be gathered with accuracy and proper presentation. System modeling is a *must* technique. The extent of system representation, choosing the swing and infinite bus, restrictions in terms of nodes (buses) and branches (lines and transformers), balanced three-phase network and a single-phase equivalent network, single line diagram, impedance diagram, etc, are all important inputs to a meaningful system study.

Section 4 deals with system modeling and data requirements to illustrate how these basic inputs for a study can be properly prepared or organized.

Once the basic preparations are completed, the next step is to look for an actual computer program, whether it be in-house or a commercial computing service. Today, many types of computing services are available, such as batch, time-sharing, and consulting services. A plant engineer must select the most suitable configuration for his needs.

Section 13 discusses the basic computation methods, various types of computer systems and their requirements, and the availability of commerical computing services and their capabilities. Section 13 gives a plant engineer the basic knowledge and direction of approach whenever he is called upon to perform a power system study with the aid of a computer program.

1.4.3 The Most Important System Studies. For the plant engineer, the following sections represent the most needed studies for the design or operation of an industrial or commercial power system: Section 5, Load Flow; Section 6, Short Circuit; Section 7, Stability; Section 8, Motor Starting; Section 9, Harmonics; Section 10, Switching Transients; Section 11, Reliability; and Section 12, Ground Mat. Each of the above sections contains a sample study which includes a computer input data file and a computer print-out for that specific study. The purpose of each study and what can be achieved by it are briefly explained.

After studying these sections, a plant engineer should be better equipped in preparing necessary data and criteria for a specific computer study if the necessity arises. The study can be performed in-house or by an outside consultant. There is a growing number of consulting firms that specialize in system studies with reasonable costs if the plant engineer can supply the necessary system data with a fair degree of correctness and accuracy.

1.5 Standard References. The following standards were used as references in the preparation of this standard:

ANSI/IEEE Std 142-1972, IEEE Recommended Practice for Grounding of Industrial and Commercial Power Systems

IEEE Std 141-1976, IEEE Recommended Practice for Electric Power Distribution for Industrial Plants

IEEE Std 241-1974, IEEE Recommended Practice for Electric Power Systems in Commercial Buildings

IEEE Std 242-1975, IEEE Recommended Practice for Protection and Coordination of Industrial and Commercial Power Systems

2. Applications of
Power System Analysis

2.1 Introduction. The planning, design and operation of industrial power systems require several studies to evaluate the current system performance, reliability, safety, and ability to grow with production requirements. Plant management is concerned with system outages, their effect on production, and in maximizing utilization of capital investment. Management requests justifications before decisions are made regarding the addition of equipment in the system and its impact on future system operation. The complexity of modern industrial power systems makes such studies difficult, tedious and time consuming to perform manually. By using digital computers, however, these computational tasks have been extremely simplified.

Modern Digital Computers. The modern digital computer offers power system engineers a powerful tool to perform more effective studies of any industrial power system. Computers help engineers design optimum power systems at minimum cost, regardless of system complexity. Recent advances in computer technology, and in particular, the intro-

duction of time-sharing techniques, have helped to reduce computing costs. Time-sharing terminals can be installed in any office. Some are portable and can be carried to the industrial site. By means of a telephone line, the terminal can be connected to a large computer system located many miles away. By way of the terminal, the engineer has access to the computer and the power system analysis programs stored in its memory. The user does not have to be familiar with computer programming or languages. Program manuals describe to the engineer, in a simple, logical way, how to feed information into the computer, what kind of input formats to use, the output options and how to store his master file for further system studies, changes and modifications. The cost of running these programs and the speedy way in which the results are obtained is very surprising. In fact, the engineer can converse with the computer to change or modify the system to meet whatever design criteria are imposed. Power system studies given below can be performed digitally using the time-sharing computer access method.

2.2 Load Flow Studies.

Load flow studies determine the voltage, current, power, and power factor or reactive power at various points in a power system under existing or contemplated conditions of normal and emergency operation. Load flow studies are essential in planning the system. They are used to determine the best operating procedure for the system, especially in the event of a loss of one or more generating units or transmission lines. They offer valuable information regarding system losses, data for equipment specifications, overall system capability and limitations, proper settings of transformers in the system, and optimization of circuit usage in the system. Load flow studies can be used to study an industrial power system under suddenly applied loads (impact loads). Impact load studies are essential in determining whether this type of load can be imposed on an existing system design. Any changes needed in system design to ensure the successful operation of the overall system are also determined. Load flow studies also determine the most suitable location of capacitors in the system for power factor improvement. Because of the complexity of load flow calculations, digital computers are used extensively in such studies. Digital computer models have been developed to study practically any industrial power system under any loading condition and to provide the initial system data (initial system power flows, voltages at various system nodes, and initial machines' electrical angles) for transient stability studies.

2.3 Fault and Short-Circuit Studies.

The current which flows in different parts of an industrial power system immediately after a fault differs from that flowing a few cycles later, just before circuit breakers are called upon to open the line on both sides of the fault. Both these currents differ widely from the current that would flow under steady-state conditions if the fault were not isolated from the rest of the system by the operation of circuit breakers. The proper selection of circuit breakers and switchgear depends on two factors, the current flowing immediately after the fault occurs and the current that the breaker must interrupt. Short-circuit calculations consist of determining these currents for various types of faults at various locations in the system. The data obtained from fault calculations also determine the settings of relays which control the circuit breakers.

Digital computers are used effectively in the calculation of the three-phase short-circuit levels at various points in the system, and in determining the momentary and interrupting ratings of circuit breakers. Line-to-ground faults and double-line short circuits can also be digitally analyzed. In this aspect, the method of symmetrical components used in the computer model represents the system sequence networks and enables the calculation of unsymmetrical faults in the system. Without digital computers, such calculations are tedious and time consuming.

2.4 Stability Studies.

In industrial power systems, the current which flows in an ac generator or synchronous motor depends on:

(1) The magnitude of its generated, or internal voltage

(2) The phase angle of its internal voltage with respect to the phase angle of the internal voltage of every other machine in the system

(3) The characteristics of the network and loads

The problem of stability is maintaining the synchronous operation of generators and motors in the system.

Stability studies are usually complex, but by using digital computers, such studies can be performed effectively at a relatively reasonable cost. Depending on the program used, the synchronous generators are simulated to include voltage regulators, excitation systems and governor response. Motors are represented by their dynamic characteristics and system disturbances are simulated digitally. Through analysis, a thorough understanding of the power system performance under various system disturbances can be determined.

2.5 Motor Starting Studies. Starting large synchronous or induction motors in an industrial power system can be a problem. It is preferable to start these motors across the line, if possible. This can cause severe voltage dips in the system however, and, in certain circumstances, the motor may not be able to *break away* from standstill or might stall during acceleration.

By using digital computer techniques, all such problems can be predicted in advance before the installation of the motor in the system. If a starting device is needed, its characteristics and rating are determined by this computer motor start-up simulation technique.

Available computer programs simulate the dynamic characteristics of the motor digitally. From these programs, the motor's voltage, current, speed, torque, and acceleration time can be computed as well as system voltage dips and currents at various time intervals from motor starting to full load speed. If a reactor or other starting device is required for motor starting, the speed at which switching should occur can be determined.

2.6 System Transients Studies. Industrial power systems that include arc furnaces and capacitor banks, resonance and switching transients can impose serious problems. By using modern digital analysis techniques, all such problems can be detected early in the design stage. These computer programs are also used in investigating the effect of adding capacitor banks to an existing system, the switching surges in a system, system response to lightning surges, lightning arrestor applications, ferro-resonance problems and system transient recovery voltage.

2.7 Reliability Analysis. When comparing various industrial power system design alternatives, system reliability and cost are essential factors in selecting optimum design. Using probability and statistical analysis techniques, the system reliability can be studied in depth with digital simulation techniques.

Interruptions. Reliability is expressed as the frequency of interruptions and expected number of hours of interruption during one year of system operation. Momentary and sustained system interruptions, component failures and outage rates are used in available digital computer reliability programs to compute overall system reliability indexes at any node in the system, and to investigate their sensitivity to parameter changes. With these results, management can select the optimum power system design and justify the decision when requested.

2.8 Power Generation Planning. In planning industrial power systems, management is faced with certain difficult questions. One problem is the decision of whether to build a local power plant or to purchase the required power from a neighboring utility system. In case of local generation expansion, the unit type, size, and time of availability in the system is another problem. By using modern digital computer analysis techniques, a thorough investigation of all such problems can be achieved.

Digitally Simulated Loads. The industrial load is simulated digitally. For specific unit outage rates, the loss of load probability associated with having insufficient generation to meet the load is calculated. Whether power is purchased or generated locally, load carrying capability of the system for various specified and tolerable loss of load probabilities in the system is then computed. The economics of both generation systems are then investigated. Depending on the type of units used in local generation (for example, diesel driven, gas turbine or fossil unit), the fuel costs, operation and maintenance costs and capacity factors are computed. Demand charges on purchased power plus other operating parameters are calculated to arrive at a comparative cost for the cases under investigation. The digital analysis model proceeds to use present worth mathematics to plant investment costs to determine the present worth costs. Yearly and total plant life fixed charges on investment, plus other fixed costs associated with the type of generation, are computed to find a final plant cost for comparison purposes. Management can now justify not only local generation or purchased power schemes, but also the type of units to be used in system expansion.

3. Analytical Procedures

3.1 Introduction. With the development of the digital computer and advanced computer programming techniques, power system problems of the most complex types can be rigorously analyzed. Previously solutions were usually only approximate and errors were introduced by many simplifying assumptions necessary to permit classical longhand calculating procedures. For progress to be realized in using the computer for power system analysis work, it has been necessary for the specialist involved in the creation of power system analysis computer programs to understand thoroughly the application of basic analytical solution methods that apply. It is also important for those concerned with assembling and preparing data for input to a power system analysis computer program and those interpreting and applying results generated by such a program to understand the application of analytical solution methods.

This section attempts, first, to identify and document the basic analytical solution methods that are valid for determining the voltage and current relationships which exist during various power system network events and operating conditions. Secondly, these basic analytical solution methods will be demonstrated where not otherwise self-evident. Finally, critical restraints that must be respected to avoid serious error in applying analytical solution methods will be discussed.

Whether a power system analysis problem is to be solved directly or by a computer program, proper application of sound analytical solution methods is essential for three reasons. First, accuracy of the solution to each individual problem being considered will be directly affected. Second, and perhaps the most important because of the significant expense involved, accuracy of the solution determines the validity and effectiveness of any remedial measures suggested. Finally, extension of erroneous results to related problems or to what appears to be a trivial modification of the original problem, possibly in combination with other misapplied or misunderstood techniques, can lead to a compounding of initial error and a progression of incorrect conclusions.

The most common causes of errors in circuit analysis work are:

(1) Failure to use a valid analytical procedure because the analyst is unaware of its existence or applicability, or both

(2) Careless or improper use of *cookbook* methods that have neither a factual basis, nor support in the technical literature, nor a valid place in electrical engineering discipline

(3) Improper use of a valid solution method due to application beyond limiting boundary restraints or in combination with an inaccurate simplifying assumption

Many situations occur in industrial and commercial power systems that illustrate some or all of these common causes of error, as well as the resulting evils. Any problem investigated as a part of the general types of power-system analysis studies covered in other sections of this recommended practice and described as follows would qualify.

(1) Short circuit studies
(2) Load analysis studies
(3) Load flow studies
(4) Stability studies
(5) Motor starting studies
(6) Harmonic studies
(7) Reliability studies
(8) Grounding mat studies
(9) Switching transient studies

3.2 The Fundamentals. The following identify the more important analytical solution methods that are either available as or are the basis for valid techniques in solving power system network circuit problems.

(1) Linearity
(2) Superposition
(3) The Thevenin Equivalent Circuit
(4) The Sinusoidal Forcing Function
(5) The Phasor Representation
(6) The Fourier Representation
(7) The Single-Phase Equivalent Circuit

(8) The Symmetrical Component Analysis

(9) The Per Unit Method

Rigorous treatment of these analytical techniques is available in several circuit analysis tests [1][1], [2], [3], [4], [5], [6], and is beyond the scope of this discussion. In the following sections, however, a brief qualitative explanation of each principle is presented, along with a review of major benefits and restraints associated with the use of each principle.

3.2.1 Linearity. Probably the simplest concept of all, linearity is also one of the most important because of its influence on the other principles. Linearity is best understood by examination of Fig 1.

Fig 1
Linearity

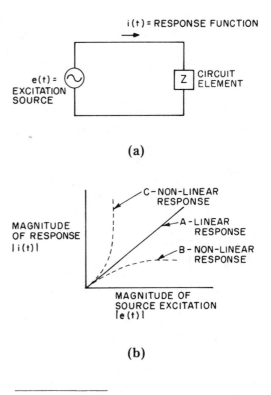

(a)

(b)

---

[1]Numbers in brackets correspond to those in the References at the end of each section.

The simplified network represented by the single-impedance element Z in Fig 1(a) is linear for the chosen excitation and response functions, if a plot of response magnitude (current) versus source excitation magnitude (voltage) is a straight line.

This is the situation shown for case A (solid line) in Fig 1(b). When linearity exists, the plot applies either to steady state value of the excitation and response functions or to the instantaneous value of the functions at a specific time.

When linear dc circuits are involved, the current doubles if the voltage is doubled. The same holds for linear ac circuits if the frequency of the driving voltage is held constant. In a similar manner, it is possible to predict the response of a *constant* impedance circuit (that is, constant R, L, and C elements) to any magnitude of dc source excitation or *fixed frequency* sinusoidal excitation based on the known response at any other level of excitation. For the chosen excitation function of voltage and the chosen response function of current, both dotted curves B and C are examples of the response characteristic of a nonlinear element.

With the circuit element represented by any of the response curves shown in Fig 1 (including the linear element depicted by curve A) the circuit will, in general, become nonlinear for a different response function, for example, power. If, for example, the element was a constant resistance (which would have a linear voltage-current relationship), the power dissipated would increase by a factor of 4 if voltage were doubled ($P = I^2R$).

An important limitation of linearity, therefore, is that it applies only to responses that are linear for the circuit conditions described (that is, a constant impedance circuit will yield a current that is linear with voltage). This restraint must be recognized in addition to the previously mentioned limitations of constant source excitation frequency for ac circuits and constant circuit element impedances for ac or dc circuits. Excitation sources, if not independent, must be linearly dependent. This restraint forces a source to behave just as would a linear response (which, by definition, is also linearly dependent).

3.2.2 Superposition. This very powerful principle is a direct consequence of linearity and can be stated as follows:

In any linear network containing several dc or fixed frequency ac excitation sources (that is, voltages), the total response (that is, current) can be calculated by algebraically adding all the individual responses caused by each independent source acting alone (that is, voltage sources shorted; current sources opened).

An example which illustrates this principle is shown in Fig 2. The equation written is for the sum of the currents from each individual source V_1 and V_2. Although Fig 2 also illustrates a way this principle might actually be used, more often its main application is in support of other calculating methods. The only restraint associated with superposition is that the network should be linear. All limitations associated with linearity apply.

The nonapplicability of superposition is why all but the very simplest nonlinear circuits are almost impossible to analyze using hand calculations. Although most real circuit elements are nonlinear to some extent, they can often be accurately represented by a linear approximation. Solutions to network problems involving such elements can be readily obtained.

Problems involving complex networks having substantially nonlinear elements

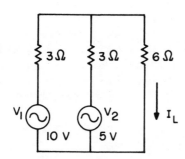

$$I_L = i_{v_1} + i_{v_2}$$

$$= \frac{10}{\left(3+\frac{6\cdot3}{6+3}\right)}\left(\frac{6\cdot3}{6+3}\right)\cdot\frac{1}{6} + \frac{5}{\left(3+\frac{6\cdot3}{6+3}\right)}\left(\frac{6\cdot3}{6+3}\right)\cdot\frac{1}{6}$$

$$= \frac{10}{5}\cdot2\cdot\frac{1}{6} + \frac{5}{5}\cdot2\cdot\frac{1}{6}$$

$$= \frac{2\cdot2}{6} + \frac{2}{6} = \frac{2}{3} + \frac{1}{3} = 1\ A$$

Fig 2
Superposition

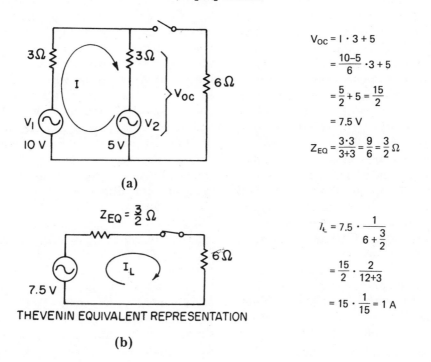

$$V_{OC} = I\cdot3 + 5$$

$$= \frac{10-5}{6}\cdot3 + 5$$

$$= \frac{5}{2} + 5 = \frac{15}{2}$$

$$= 7.5\ V$$

$$Z_{EQ} = \frac{3\cdot3}{3+3} = \frac{9}{6} = \frac{3}{2}\ \Omega$$

(a)

$$I_L = 7.5\cdot\frac{1}{6+\frac{3}{2}}$$

$$= \frac{15}{2}\cdot\frac{2}{12+3}$$

$$= 15\cdot\frac{1}{15} = 1\ A$$

THEVENIN EQUIVALENT REPRESENTATION

(b)

Fig 3
The Thevenin Equivalent

31

can practically be solved only through the use of certain simplification procedures, or through adjustment of calculated results to correct for nonlinearity. Both of these approaches can potentially lead to significant inaccuracy. Tiresome iterative calculations performed in an instant by the digital computer make accurate solutions possible when an equation can be written mathematically to describe the nonlinear circuit elements.

3.2.3 The Thevenin Equivalent Circuit. This powerful circuit analysis tool is based on the fact that any active linear network, however complex, can be represented by a single voltage source equal to the open-circuit voltage across any two terminals of interest, in series with the equivalent impedance of the network viewed from the same two terminals with all sources in the network inactivated (that is, voltage sources shorted and current sources opened). Validity of this representation requires only that the network be linear. Existence of linearity is a necessary restraint. Application of the Thevenin equivalent circuit can be appreciated by referring to the simple circuit of Fig 2 and developing the Thevenin equivalent for the network with the switch in the open position as illustrated in Fig 3. After connecting the 6 Ω load to the Thevenin equivalent network by closing the switch, the solution for I_L is the same as before, 1 A. Use of the simple Thevenin equivalent shown for the entire left side of the network makes it easy to examine circuit response as the load impedance value is varied.

The Thevenin equivalent circuit solution method is equally valid for complex impedance circuits. It is the type of representation shown in Fig 3 that is the basis for making per unit short-circuit calculations, although the actual values

for the source voltage and branch impedances would be substantially different from those used in this case. (The circuit property of linearity would, incidentally, allow them to be scaled up or down.) The network shown in Fig 3(a), with the 6 Ω resistance shorted and the other resistances visualized as reactances, might well serve as an oversimplified representation of a power system about to experience a bolted fault with the closing of the switch.

The V_1 branch of the circuit would correspond to the utility supply while the V_2 branch might represent a large motor running unloaded, immediately adjacent to the fault bus, and highly idealized so as to have no rotor flux leakage. For such a model, the 5 V source corresponds to the pre-fault, air-gap voltage behind a stator leakage (subtransient) reactance of 3 Ω [7]. In a more realistic situation where rotor leakage is evident, a model that accurately describes the V_2 branch in detail before and after switch closing is much more difficult to develop, because the air-gap voltage decreases (exponentially) with time and varies (linearly) with the steady-state rms magnitude of the motor stator current following application of the fault. The problem of accounting for motor internal behavior is avoided altogether by use of a Thevenin equivalent. This permits the V_2 branch to be represented by the apparent motor reactance (or, more generally, impedance) effective at the time following switch closure. In shunt with the equivalent impedance for the remainder of the network, the Thevenin equivalent impedance for the motor (at any point in time of interest) is simply connected in series with the pre-fault open-circuit voltage to obtain the corresponding current response to switch closing.

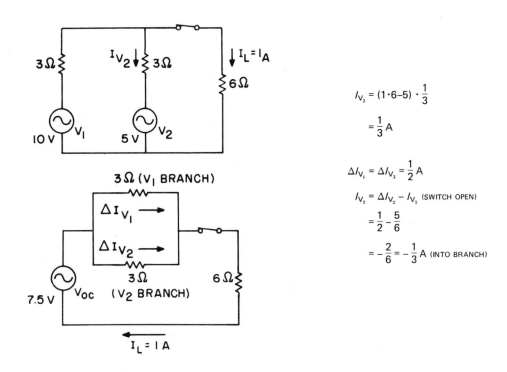

$$I_{V_2} = (1 \cdot 6 - 5) \cdot \frac{1}{3}$$

$$= \frac{1}{3} A$$

$$\Delta I_{V_1} = \Delta I_{V_2} = \frac{1}{2} A$$

$$I_{V_2} = \Delta I_{V_2} - I_{V_2} \text{ (SWITCH OPEN)}$$

$$= \frac{1}{2} - \frac{5}{6}$$

$$= -\frac{2}{6} = -\frac{1}{3} A \text{ (INTO BRANCH)}$$

Fig 4
Current Flow of a Thevenin
Equivalent Representation

The current response obtained in each branch of a network using a Thevenin equivalent circuit solution represents the change of current in that branch. The actual current that flows is the vector sum of currents before and after the particular switching event being considered. See Fig 4.

In Fig 4A the current flowing in the V_2 branch circuit is shown to be $1/3$ A. A more detailed representation of the Thevenin equivalent circuit previously examined in Fig 3 is shown in Fig 4B. Here, the solution for the same current I_{V2} is determined by subtracting the current flowing in the V_2 branch prior to closing the switch ($5/6$ A from inspection of the circuit in Fig 3A) from the current $I_{V2} = 1/2$ A, calculated to be flowing in the Thevenin equivalent for this V_2 branch.

In the branch of the circuit defined by the switch itself, the *change of current* due to closing is normally the response of interest. This means the solution to the Thevenin equivalent is sufficient. The resultant current in the other branches, however, cannot be determined by the solution to the Thevenin equivalent network alone.

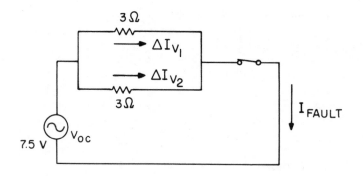

$$I_{FAULT} = \frac{15}{2} \cdot \frac{2}{3} = 5 \text{ A}$$

$$\Delta I_{V_1} = \Delta I_{V_2} = 2.5 \text{ A} = \frac{5}{2} = \frac{15}{6}$$

$$I_{V_2} = \Delta I_{V_2} - I_{V_2} \text{ (SWITCH OPEN)}$$

$$I_{V_2} = \frac{15}{6} - \frac{5}{6} = \frac{10}{6} \text{ A}$$

Fig 5
Fault Flow

In the case where the V_2 branch represents a motor switched onto a bolted fault, the motor contribution is the locked-rotor current minus the pre-fault current as illustrated in Fig 5 and not just the locked-rotor current as it is so often carelessly described. As a rule, this effect is never as significant as the example suggests, even when the motor is loaded prior to the fault; the load current is much smaller than the locked-rotor current and almost 90° out of phase with it.

A Norton equivalent which consists of a current source in parallel with a (different) equivalent impedance can alternately be developed for the Thevenin equivalent circuit. This representation is not generally as useful in power system analysis work.

3.2.4 The Sinusoidal Forcing Function. It is a most fortunate truth in nature that the excitation sources (that is, driving voltage) for electrical networks, in general, have a sinusoidal character and can be represented by a sine wave plot of the type illustrated in Fig 6. There are two important consequences of this circumstance. First, although the response, that is, current, for a complex R, L, C network represents the solution to at least one second-order differential equation, the result will also be a sinusoid of the same frequency as the excitation and different only in magnitude and phase angle. The relative character of the current with respect to the voltage for simple R, L, and C circuits is also shown in Fig 6.

The second important concept is that

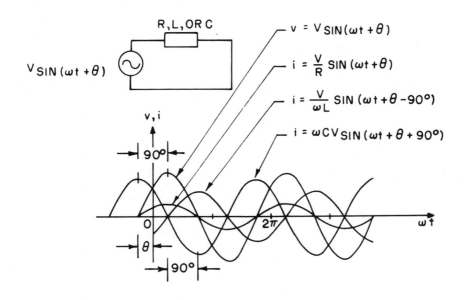

Fig 6
The Sinusoidal Forcing Function

when the sine wave shape of current is forced to flow in a general impedance network of R, L, and C elements, the voltage drop across each element will always exhibit a sinusoidal shape of the same frequency as the source. The sinusoidal character of all the circuit responses makes the application of the superposition technique to a network with multiple sources surprisingly manageable. The necessary manipulation of the sinusoidal terms is easily accomplished using the laws of vector algebra, which evolve from the next technique to be reviewed.

The only restraint associated with the use of the sinusoidal forcing function concept is that the circuit must be comprised of linear elements, that is, R, L, and C are constant as current or voltage varies.

3.2.5 Phasor Representation. Phasor representation allows any sinusoidal forcing function to be represented as a phasor in a complex coordinate system as shown in Fig 7. As indicated, the expression for the phasor representation of a sinusoid can assume any of the following shorthand forms:

Exponential: $E\,e^{j\theta}$
Rectangular: $E \cos \theta + jE \sin \theta$
Polar: $E\,\underline{/\theta}$

For most calculations, it is more convenient to work in the *frequency domain* where any angular velocity associated with the phasor is ignored, which is equivalent to assuming the coordinate system rotates at a constant angular velocity of ω.

The impedances of the network can likewise be represented as phasors using the vectorial relationships shown. As illustrated, the circuit responses, that is,

$$E\, e^{j\theta} = E \cos \theta + jE \sin \theta = E\, \underline{/\theta}$$

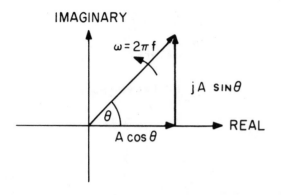

IMAGINARY

$\omega = 2\pi f$

$jA\ \sin\theta$

θ

REAL

$A \cos \theta$

$$I = \frac{E}{Z} = \frac{|E|}{|Z|}\frac{e^{j\theta}}{e^{j\phi}} = \frac{|E|}{|Z|}\, e^{j(\theta - \phi)}$$

$$Z = R + j\,\omega L - j\,\frac{1}{\omega C} = R + j\,X$$

$$X_L = \omega L$$

$$X_C = \frac{1}{\omega C}$$

$$|Z| = +\sqrt{R^2 + X^2}$$

$$\phi = \tan^{-1}\frac{X}{R}$$

R = resistance
X_L = inductive reactance
X_C = capacitive reactance
Z = impedance

**Fig 7
The Phasor Representation**

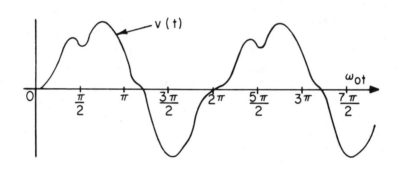

v (t)

$\omega_0 t$

0    $\dfrac{\pi}{2}$ π $\dfrac{3\pi}{2}$ 2π $\dfrac{5\pi}{2}$ 3π $\dfrac{7\pi}{2}$

$$V(t) = V_0 + V_1 \cos \omega_0 t + V_2 \cos 2\omega_0 t + \ldots$$

$$+ V_1' \sin \omega_0 t + V_2' \sin 2\omega_0 t + \ldots$$

**Fig 8
The Fourier Representation**

current, can be obtained through the simple vector algebraic manipulation of the quantities involved. The need for solving complex differential equations to determine the circuit responses is completely eliminated.

The restraints that apply are:

(1) The sources must all be sinusoidal

(2) The frequency must remain constant

(3) The circuit R, L and C elements must remain constant (that is, linearity must exist)

3.2.6 The Fourier Representation.
This powerful tool allows any nonsinusoidal periodic forcing function, of the type plotted in Fig 8, to be represented as the sum of a dc component and a series (infinitely long, if necessary) of ac sinusoidal forcing functions. The ac components have frequencies that are an integral *harmonic* of the periodic function (fundamental) frequency. The general mathematical form of the so-called *Fourier series* is also shown in Fig 8.

The importance of the Fourier representation is immediately apparent. The response to the original driving function can be determined by first solving for the response to each Fourier series component forcing function and summing all the individual solutions to find the total superposition. Since each component response solution is readily obtained, the most difficult part of the problem becomes the determination of the component forcing function. The individual harmonic voltages can be obtained, occasionally in combination with numerical integration approximating techniques through several well-established mathematical procedures. Detailed discussion of their use is better reserved for the many excellent texts [2], [3] that treat the subject.

There are several abstract mathematical conditions that must be satisfied to use a Fourier representation. The only restraints of practical interest to the power systems analyst are that the original driving function must be periodic (repeating) and the network must remain linear.

3.2.7 The Single-Phase Equivalent Circuit.
The single-phase equivalent circuit is a powerful tool for simplifying the analysis of balanced three-phase circuits, yet its restraints are probably most often disregarded. Its application is best understood by examining a three-phase diagram of a simple system and its single-phase equivalent, as shown in Fig 9. Also illustrated is the popular one-line diagram representation commonly used to describe the same three-phase system on engineering drawings.

If a three-phase system has a perfectly balanced symmetrical source excitation (voltage) and load, as well as equal series and shunt system and line impedances connected to all three phases (see Fig 9(a)), imagine a conductor (shown dotted) carrying no current connected between the effective neutrals of the load and the source. Under these conditions, the system can be accurately described by either Fig 9(b) or Fig 9(c).

The single-phase equivalent circuit is particularly useful since the solution to the classical loop equations is much easier to obtain than for the more complicated three-phase network. To determine the complete solution, it is only necessary to realize that the other two phases will have responses that are shifted by 120° and 240° but are otherwise identical to the reference phase.

Anything that upsets the balance of the network renders the model invalid. A subtle way this might occur is illustrated in Fig 10. If the switching devices operate independently in each of the three

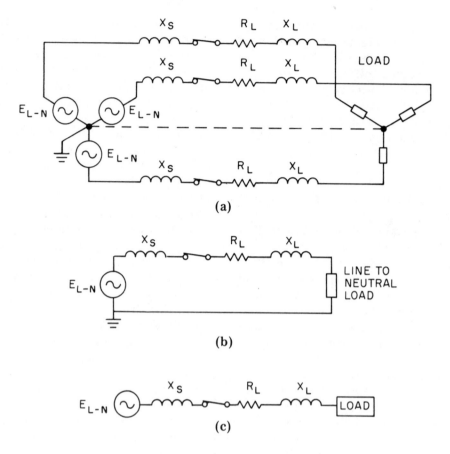

Fig 9
(a) Three-Phase Diagram, (b) Single-Phase
Equivalent, and (c) One-Line Diagram

poles, and for some reason the device in phase A becomes opened, the balance or symmetry of the circuit is destroyed. Neither the single-phase equivalent nor the one-line diagram representation is valid. Even though the single-phase and the one-line diagram representations would imply that the load has been disconnected, it continues to be energized by single-phase power. This can cause serious damage to motors and result in unacceptable operation of certain load apparatus.

More importantly, if only one switch-

ing device operates in response to a fault conditon in the same phase, as depicted at location X, the system sources would continue to supply fault current from the other unopened phases through the impedance of the load. The throttling effect of the normally substantial load impedance, possibly in combination with additional arc impedance, can reduce the level of the current to a point where detection may not occur in phases (b) and (c). Needless to say, substantial damage can result before the fault finally burns enough to involve the other phases

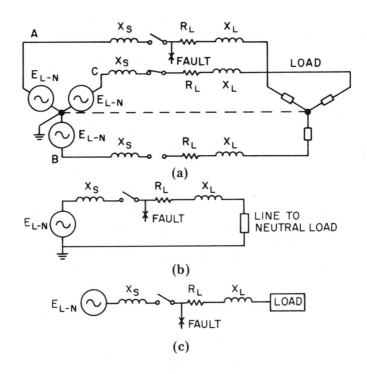

Fig 10
(a) Three-Phase Diagram, (b) Single-Phase
Equivalent, and (c) One-Line Diagram

directly and accomplish complete inter-ruption. Meanwhile, both of the single-line representations fail to recognize the problem, and in fact, suggest that the condition has been safely disconnected. Therefore, the restraints of this calculating aid are:

(1) Symmetry of the electrical system, including all switching devices, and balance of the applied load

(2) Any of the other previously described restraints which apply to the analytical technique being used in combination with the single-phase equivalent

3.2.8 The Symmetrical Component Analysis. This approach comes to the analyst's rescue when he is confronted with an unbalance, the most common circuit condition which invalidates the single-phase equivalent circuit solution method. The symmetrical component analysis allows the response to any unbalanced condition in a three-phase power system to be investigated and correctly synthesized by the sum of the responses to as many as three separate balanced system conditions. The application of an unbalanced set of voltage phasors, such as displayed in Fig 11, to a balanced downstream system and load is the sum of the responses of the balanced components which vectorially add to form the original unbalanced set.

Similar conclusions apply when the voltages are balanced, but the connected phase impedances or loads, or both, and therefore, the line currents are unbalanced. Here, the unbalanced current phasors are the sum of up to three balanced sets that flow through the balanced

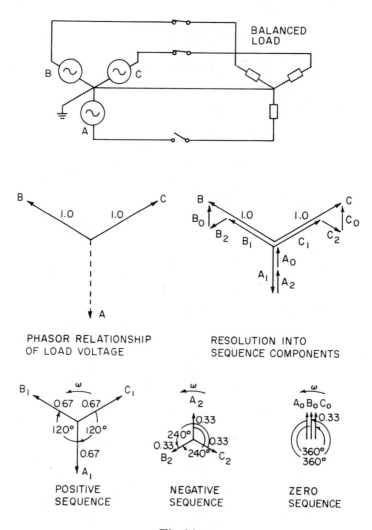

PHASOR RELATIONSHIP
OF LOAD VOLTAGE

RESOLUTION INTO
SEQUENCE COMPONENTS

POSITIVE
SEQUENCE

NEGATIVE
SEQUENCE

ZERO
SEQUENCE

Fig 11
The Symmetrical Component Analysis

system impedances on either or both sides of the unbalance producing voltage drops that satisfy the needs of the applied voltages and the boundary conditions at the point of unbalance.

The mathematical expression for three unbalanced phasors (see Fig 11) as a function of the balanced phasor components is as follows:

$$\overline{A} = \overline{A}_0 + \overline{A}_1 + \overline{A}_2$$

$$\overline{B} = \overline{A}_0 + a^2\overline{A}_1 + a\overline{A}_2$$

$$\overline{C} = \overline{A}_0 + a\overline{A}_1 + a^2\overline{A}_2$$

The positive (1), negative (2), and zero (0) sequence vector components of any phase always have the angular relationship with respect to one another as described in the vector diagram of Fig 11.

These are assigned a counter-clockwise direction of rotation in the *time domain* as illustrated. In the *space domain*, the negative sequence phasors will produce exactly the same results as a set of equal magnitude phasors that are displaced from one another by 120° and that rotate clockwise with time.

Although the proof that a set of N unbalanced vectors can be completely represented by N sets of balanced vectors is seldom presented in the texts dealing with the subject of symmetrical components, it is true. First, it is postulated that it might be possible to describe three arbitrary but defined vectors, $\overline{A}, \overline{B},$ and $\overline{C}$ by the expressions (previously provided) involving only the three unknowns $\overline{A}_0$, $\overline{A}_1$, and $\overline{A}_2$. Then the terms are rearranged to solve for these unknowns as follows:

$$\overline{A}_0 = \frac{1}{3} (\overline{A} + \overline{B} + \overline{C})$$

$$\overline{A}_1 = \frac{1}{3} (\overline{A} + a\overline{B} + a^2\overline{C})$$

$$\overline{A}_2 = \frac{1}{3} (\overline{A} + a^2\overline{B} + a\overline{C})$$

Here are three independent equations for three unknowns ($\overline{A}_0$, $\overline{A}_1$ and $\overline{A}_2$), which are all that are required to uniquely and completely describe $\overline{A}_0$, $\overline{A}_1$, and $\overline{A}_2$ and, therefore, substantiate their existence. The corresponding B and C phase components are, then, defined by the vector relationship shown in Fig 11.

The merit of the symmetrical component analysis is that a relatively complicated, and often unwieldy problem, can be solved by simply vectorially summing the solution to no more than three balanced network problems. Several reference texts [7], [1], [5] provide convenient tables showing the network interconnections that must be used to solve for the responses to many commonly encountered system unbalances, as well as certain balanced conditions.

The three impedance networks (positive, negative, and zero sequence) are symbolically represented in short-hand fashion by an empty *block diagram* for the phase most definitive of the condition being studied up to the unbalance or other point of interest. Here, the final interconnections of the networks are shown which satisfy the necessary boundary conditions describing the system at the point of concern. The analyst can fill the *block diagrams* with the proper sources and impedances, including loads, in each sequence network and solve the single-phase loop equations. This, then, produces the three sequence responses, that is, current or voltage, which add vectorially to produce the resultant phase responses. Similarly, the other phase responses can be obtained by adding vectorially the individual sequence solutions shifted by the appropriate multiple of 120°.

One curious and often confounding feature of this solution procedure is that the phase in the system which usually provides the best, and sometimes the only, approach to the solution for an unbalance is the one least actively involved in the event. The unbalance illustrated in Fig 11 is one such example where the solution is obtained through an analysis of the non conducting phase A. A double line-to-ground fault is another, where examination of the open phase gives the most direct access to the network solution.

The symmetrical component analysis always involves the use of superposition as well as most of the other procedures previously discussed. The restraints which apply to these other procedures, therefore, must also govern the use of the symmetrical component analysis. In addition, due to mutual phase winding coupling and other effects, the impedance displayed by electrical machines will be different when excited by the different sequence sources. Hence, the *per phase* impedance of the negative and zero sequence networks will, in general, be different from the positive. Currents flowing in the zero sequence network, being in phase, do not add to zero as do both the positive and negative sequence currents and are, therefore, influenced by impedance in this additional circuit path. When harmonic excitation sources are present (requiring the use of the Fourier representation) special care must be exercised in treating the sequence networks. Starting with the fundamental, the harmonic term progressively shifts from the positive to the negative to the zero sequence networks, and then the process repeats.

3.2.9 The Per Unit Method. This method of calculation and its close companion, the percentage method, are well documented in [4], [5] and [6] and are generally well known. As a result, they will only be mentioned in passing here.

Fundamentally, the per unit method and the percentage method amount to a short-hand calculating procedure where all equivalent system and circuit impedances are converted to a common kVA *base*. This permits the ready combination of circuit elements in a network where different system voltages are present without the need to convert impedances each time responses are to be determined at a different voltage level.

Associated with each impedance element and its kVA base is a line-to-line kV base (usually the *nominal* line voltage at which the element is connected to the system), along with the resulting *base impedance*, and *base current* related by the following expressions:

(1) *Three-Phase Network*

$$I_{base} = \frac{kVA_{base}}{\sqrt{3}\ kV_{base}}$$

$$Z_{base} = \frac{kV_{base}^2}{kVA_{base}} \cdot 1000$$

$$kVA_{base} = \sqrt{3}\ kV_{base}\ I_{base}$$

(2) *Single-Phase Network*

$$I_{base} = \frac{kVA_{base}}{kV_{base}}$$

$$Z_{base} = \frac{kV_{base}^2}{kVA_{base}} \cdot 1000$$

$$kVA_{base} = kV_{base}\ I_{base}$$

The equivalent open-circuit driving voltage ahead of the entire network can then conveniently and properly be expressed as 1.0 per unit (or 100%) if the base kV is selected to match the operating voltage. The method is simply illustrated and compared to the conventional ohmic calculation method for a single phase network in Fig 12 where the equivalent circuit impedances as viewed from the primary and secondary sides at the transformer are identified as Z_p and Z_s, respectively.

Before combining impedances it is essential that the corresponding per unit, or percentage, ratings have been expressed on a common kVA base and that they are connected to a system having a line-to-line voltage equal to the kV base to which the per unit, or percentage, rating is referenced. Since the per unit method of calculation is based on the existence of linearity and is always used in combination with one or more of the other principles, it is necessary to observe all of the associated restraints discussed earlier as they apply.

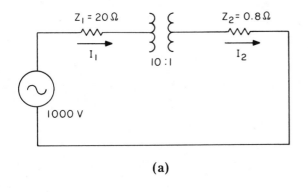

$$Z_P = 20 + \left(\frac{10}{1}\right)^2 \cdot 0.8 = 100 \ \Omega$$

$$Z_S = 20 \cdot \left(\frac{1}{10}\right)^2 + 0.8 = 1.0 \ \Omega$$

$$I_1 = \frac{1000}{100} = 10 \ A$$

$$I_2 = 1000 \cdot \left(\frac{1}{10}\right) \div 10 = 100 \ A$$

base $kV_P = 1$

base $kV_S = \frac{1}{10}$

base MVA = 5

$$Z_1 = 20 \div \frac{1.0^2 \cdot 1000}{5000} = 100 \text{ per unit}$$

$$Z_2 = 0.8 \div \frac{(1/10)^2 \cdot 1000}{5000} = 400 \text{ per unit}$$

$$Z_{EQ} = 100 + 400 = 500 \text{ per unit}$$

$$I_1 = I_2 = \frac{1.0 \text{ per unit volts}}{500 \text{ per unit}} = 0.002 \text{ per unit}$$

$$I_{1base} = \frac{5MVA}{1kV} = 5000 \ A$$

$$I_1 = (5000) \cdot (0.002) = 10 \ A$$

$$I_{2base} = \frac{5MVA}{0.1kV} = 50\ 000 \ A$$

$$I_2 = (50\ 000) \cdot (0.002) = 100 \ A$$

(a)

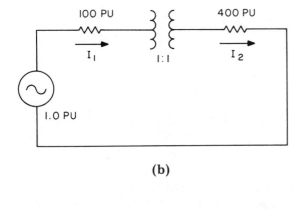

(b)

Fig 12
(a) Classical Ohmic Representation,
(b) Per Unit Representation

3.3 References and Bibliography

[1] WAGNER, C. F., and EVANS, R. D. *Symmetrical Components*, New York: McGraw-Hill, 1933.

[2] HOYT, Jr, W. H., and KEMMERLY, J. E. *Engineering Circuit Analysis*, New York: McGraw-Hill, 1962.

[3] CLOSE, C. M. *The Analysis of Linear Circuits*, New York: Harcourt, Brace and World, Inc, 1966.

[4] STEVENSON, Jr, W. D. *Elements of Power System Analysis*, New York: McGraw-Hill, 1962.

[5] BEEMAN, DONALD. *Industrial Power Systems Handbook*, New York: McGraw-Hill, 1955, chap 2.

[6] WEEDY, B. M. *Electric Power Systems*, New York: John Wiley and Sons, Inc, 1972, chap 2.

[7] *Electrical Transmission and Distribution Reference Book*, East Pittsburgh, Pa: Westinghouse Electric Corporation, 1964, chaps 2 and 6.

[8] FITZGERALD, A. E., and KINGSLEY, Jr, CHARLES. *Electric Machinery*, New York: McGraw-Hill, 1961.

[9] PUCHSTEIN, A. F., and LLOYD, T. C. *Alternating Current Machines*, New York: John Wiley and Sons, Inc, 1947.

4. System Modeling

4.1 Introduction. This section addresses the following questions:

(1) How can each component or each group of components of an industrial or commercial power system be represented so that an analysis of the system performance can be made

(2) Which of the several possible representations, or *models*, of the given components will best describe the system to meet the objectives of a given study

(3) What mathematical expressions will describe the characteristics of each system element so that it can be quantified and programmed for computer input

There are an infinite number of possible power system configurations and a large variety of study types. Consequently standards cannot be established to dictate specific models for all specific circumstances.[2] This text will therefore serve as a guide to help the reader make judicious trade-offs in selecting models for his study.

Derivations and proofs of mathematical expressions will not be given. References should be used for such purpose. However, several fundamental relationships of electrical and mechanical quantities will be mentioned in the text. This should save the reader the time needed to locate them in text books or handbooks. It should also help refresh the memories of those who have not been exposed to power studies or academic activities for a long time.

The material is intended to be as basic and simple as permitted by the subject. The emphasis is to correlate real life systems with the abstraction of mathematics in order to communicate with the computer properly.

4.2 Modeling. Scale modeling of power systems as a means of analyzing their performance is impractical. However, scale models of certain mechanical components of power systems are used to evaluate their characteristics. This is often the case with hydraulic sections of hydroelectric plants such as turbine runners, spiral cases, gates, draft tubes,

---

[2]An exception to this statement are the application standards for circuit breakers where models for sources of short circuit current are specified. See 4.9.2.2 and 4.10.3.

etc. Nonetheless, much expertise is required to establish scaling and normalization factors, to construct the model, to gather meaningful data by measurement and to interpret and extrapolate the results.

A step in the direction of abstraction is the approach used in ac and dc analyzer boards for the study of power systems. Here the power system elements are modeled by modules of equivalent characteristics but physically much smaller. Most modules do not bear visual resemblance to the components they represent. A 200 mi transmission line module may look like a few potentiometers, inductances and capacitors. The modules are adjustable so the system variables can be modeled easily. These modules can be interconnected, each adjusted to satisfy the conditions of the power system being studied, and the required measurements made and interpreted. Once the board is configured for a base case, the effect of one variable on the overall system can be readily analyzed simply by adjusting the affected module. Most analyzing boards have been abandoned today in favor of the more economical digital method using computers.

Digital computers can be programmed to solve quickly and at relatively low cost a large number of simultaneous equations and can handle the algebra of large matrices. This makes them particularly well suited for applications in power system analysis. An immense variety of programs have been written to study an ever increasing number of problems in the electrical field. These programs are usually set up to receive the problem information in the form of numbers rather than analog settings. This then forces the power system analyst to model the system quantitatively. The programs, designed to maximize their usefulness are written in general non-specific terms. This exposes the analyst to a choice of program features and alternatives that require decisions to be made every step of the way. Finally the programs are often structured to handle extensive power systems (3000 bus programs are not uncommon). This aspect suggests that the analyst should consider to what extent his system will be modeled to avoid, on the one hand, an expensive overkill and on the other an incomplete problem statement that ensures questionable answers.

4.3 Review of Basics. Power network elements may be classified in two categories, passive elements and active elements.

4.3.1 Passive Elements. The passive elements comprise such components as transmission lines, transformers, reactors, and capacitors. They will, in general, be regarded as linear. They will be modeled by one or more of the following electrical quantities:

Name	Symbol	Unit
resistance	R	ohm
inductance	L	henry
capacitance	C	farad

The voltage across and the current through the element will be governed by these relationships:

$$v = Ri \qquad i = \frac{v}{R} \qquad \text{(Eq 1)}$$

$$v = L\frac{di}{dt} \qquad i = \frac{1}{L}\int v\, dt \qquad \text{(Eq 2)}$$

$$v = \frac{1}{C}\int i\, dt \qquad i = C\frac{dv}{dt} \qquad \text{(Eq 3)}$$

where the lowercase letters represent the instantaneous values of voltage and current.

In dc circuits under steady state conditions these equations will reduce to:

$$V = RI \qquad I = \frac{V}{R}$$

$$V = 0 \quad \left(\text{since } \frac{di}{dt} = 0\right)$$

$$I = 0 \quad \left(\text{since } \frac{dv}{dt} = 0\right) \qquad \text{(Eq 4)}$$

In ac circuits with sinusoidal wave shapes, the equations become:

$$V = RI \qquad I = \frac{V}{R}$$

$$V = jX_L I$$

where

$X_L = 2\pi f L$
= inductive reactance   (Eq 5)

$$V = -jX_C I$$

where

$X_C = \dfrac{1}{2\pi f C}$
= capacitive reactance   (Eq 6)

The capital letters for voltages and currents represent their rms values, f is the frequency in hertz, and j the $90°$ operator ($= \sqrt{-1}$). Inverting and combining these elements in series or parallel will define the set of quantities of Table 1.

It should be noted here that it is customary in ac power circuits to use the R, X and Z quantities for the series (line) elements and the G, B and Y quantities for the shunt (line to neutral) elements.

Note also that Z and Y are complex quantities that can be expressed in the rectangular form above or the polar form $Z = |Z|\,\underline{/\theta}$ or $Y = |Y|\,\underline{/\theta}$. Most computer programs accept the Z and Y values in the rectangular form.

A final remark concerns the sign ahead of the reactances and susceptances. The four diagrams of Fig 13 are self explanatory. The wise analyst will verify the program instructions to make sure that the computer will interpret the input data properly.

4.3.2 Active Elements. The active elements of a power system comprise such components as motors, generators, synchronous condensers, other loads like furnaces, adjustable speed drives, etc. The active elements will be regarded as nonlinear, although some of the components may behave linearly under certain circumstances.

One or more of the parameters of a model of an active element will vary as a function of time, phase angle, frequency, speed, etc.

The four expressions for power quantities given in Table 2 can be used to model non linear elements. Given any two of the four values, the remaining two can be defined. Power can also be expressed in polar form: $S = |S|\,\underline{/\theta}$ which yields these relationships: $PF = \cos\theta$, $P = S\cos\theta$ and $Q = S\sin\theta$.

Table 1
Equation References for Conductance, Susceptance, Impedance and Admittance

Name	Symbol	Unit	Defining Equation
conductance	G	S (mho)	$1/R$
inductive susceptance	B	S (mho)	$1/X$
capacitive susceptance	B	S (mho)	$1/X$
impedance	Z	S (ohm)	$(R + jX)$
admittance	Y	S (mho)	$(G + jB)$

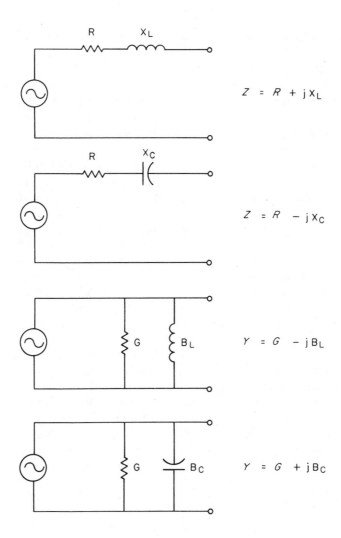

Fig 13
Squirrel Cage Induction Motor Model

Table 2
Four Expressions for Power Quantities

Name	Symbol	Unit	Expression
complex power	S	VA	$S = P + jQ$
active power	P	W	$P = \sqrt{S^2 - Q^2}$
reactive power	Q	var	$Q = \sqrt{S^2 - P^2}$
power factor	PF	per unit	$\text{PF} = \dfrac{P}{S}$

Note that the signs of P or Q may be positive or negative. By convention, the positive sign of Q is used for inductive loads, that is, the current will lag the voltage applied to a load that *consumes* vars. In this sense it is said that capacitors *generate* vars (current leads the voltage) and that squirrel cage induction motors *absorb* vars.

By convention also, the sign of P is positive for a load that consumes energy or a source that generates energy. Thus a load with a negative sign for P could be used to represent a generator, and vice versa.

Some of the power system components can best be expressed in terms of current or voltage. For instance, an infinite bus may be specified by a voltage source of constant magnitude and angle, and a particular load may be described as a constant current element. The current and voltage quantities may be complex numbers, in which case they have to be described in terms of a reference vector that may be a voltage or current quantity. This introduces the phase angle concept:

$$I = |I| \; \underline{/\theta} = I_x + jI_y = I(\cos\theta + j\sin\theta)$$
$$V = |V| \; \underline{/\theta} = V_x + jV_y = V(\cos\theta + j\sin\theta)$$

where the x axis of the coordinate system is taken as the reference shown in Fig 14.

So far we have reviewed most of the quantities used in the type of studies that require network solutions at a single instant: a *snapshot* of the network in equilibrium. This instant shows the system in a steady state or during a transient, at an instant for which all network parameters are known, say ½ cycle after a short circuit has occurred.

Some studies, of which motor starting and transient stability are examples, require that a complete period of time be covered to assess the effects of a disturbance on system performance. This requirement for a certain period of time to be considered introduces the need for mechanical quantities.

The fundamental quantities of mechanics are space, matter and time. Length measures space and mass measures matter. Other mechanical quantities are derived from these three. Some of the derived quantities officially recognized

Fig 14
Section of a Typical
Industrial Plant Impedance Diagram

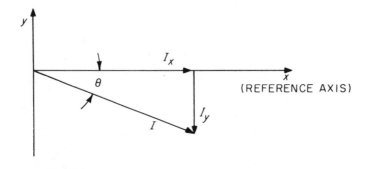

Table 3
Fundamental Equations for
Translation and Rotation

Name	Symbol	Unit	Defining Equation
Fundamental			
length	l	meter (m)	
mass	m	kilogram (kg)	
time	t	second (s)	
Translation			
velocity	v	m/s	$v = \dfrac{dl}{dt}$
acceleration	a	m/s²	$a = \dfrac{d^2 l}{dt^2}$
force	F	newton (N)	$F = ma$
work	W	joule (J)	$W = \int F\, dl$
power	P	watt (W)	$P = \dfrac{dw}{dt}$
momentum	M'	N/s	$M' = mv$
Rotation			
radius	r	m	
circular arc	s	m	
moment of inertia	I	kg-m²	$I = \int r^2 dm$
angle	θ	radian (rad)	$\theta = \dfrac{s}{r}$
angular velocity	ω	rad/s	$\omega = \dfrac{d\theta}{dt}$
angular acceleration	α	rad/s²	$\alpha = \dfrac{d^2\theta}{dt^2}$
torque	T	N·m	$T = rF$
work	W	J	$W = \int T\, d\theta$
power	P	W	$P = T\omega$
angular momentum	M	J s/rad	$M = I\omega$

for electrical engineering work have been tabulated in Table 3, with the MKS system of units and the defining equations.

4.4 Power Network Solution. Before dealing with the detailed models of power system components, it is important to review what constitutes the solution of a network.

It can be said that a network is resolved if all the bus voltages and the relative phase angles between these voltages are known. This of course requires that the impedance between the buses be known.

It also requires that a reference voltage and angle be specified for one bus.

Consider for instance Fig 15 which shows a small section of the typical plant electrical system of Section 5, Load Flow. Assume that the voltages at buses 2, 4 and 24 (also called A, B and C, to simplify notations) and the impedances of T2 and line 8 are known. They are summarized on Fig 16 and listed as follows:

$$V_2 = V_A = 69.00 \text{ kV } \underline{/0^\circ}$$
$$V_4 = V_B = 13.60 \text{ kV } \underline{/-1.6^\circ}$$
$$V_{24} = V_C = 13.51 \text{ kV } \underline{/-1.82^\circ}$$

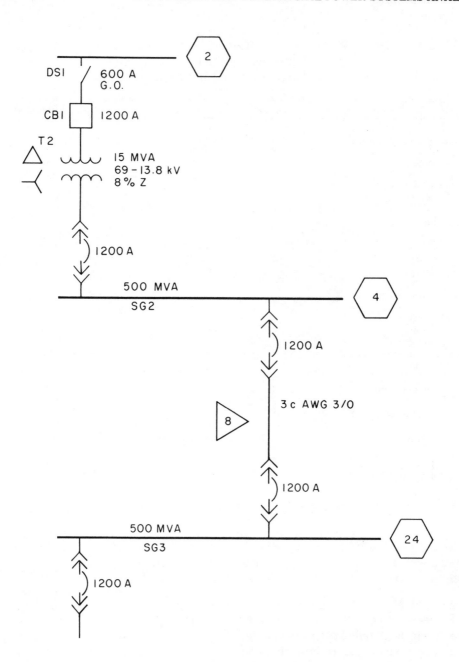

Fig 15
Single Line Diagram

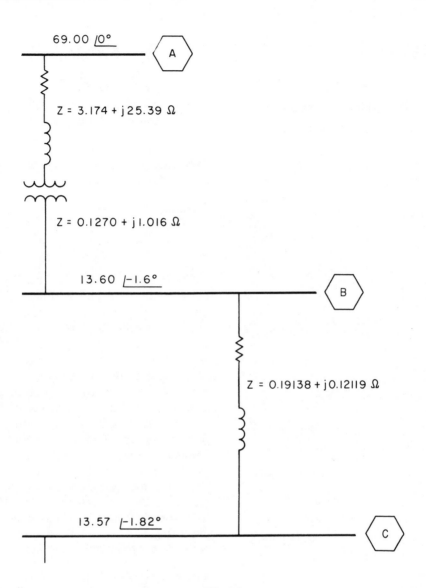

Fig 16
Impedance Diagram

Z_{T2} = Z_{AB} = 1.0 + j8.0 (%) on
15 MVA base

= $(0.01 + j0.08) \cdot \dfrac{69^2}{15}$

= 3.174 + j25.39 Ω (at 69 kV)

= $(0.01 + j0.08) \cdot \dfrac{13.8^2}{15}$

= 0.1270 + j1.016 Ω (at 13.8kV)

Z_{L8} = Z_{BC} = 0.19138 + j0.12119

The current from bus A to bus B can be found by:

I_{AB} = $(V_B - V_A)/Z_{AB}\sqrt{3}$ (Eq 7)

= 10^3 (69.00 $\underline{/0°}$ – 13.60

$\cdot \dfrac{69}{13.8}$ $\underline{/-1.6°}$)

$/(3.174 + j25.39)\sqrt{3}$

= 10^3 [69.00+ j0 – (67.97

– j1.90)]$/(25.59 \underline{/82.87°})\sqrt{3}$

= 10^3 (1.03 + j 1.90)

$/(25.59 \underline{/82.87°})\sqrt{3}$

= 10^3 (2.16$\underline{/61.54}$)

$/(25.59 \underline{/82.87°})\sqrt{3}$

= $10^3 \cdot 2.16$

$/25.59\sqrt{3}$ $\underline{/61.54 - 82.87°}$

= 48.73 $\underline{/-21.33°}$ A (at 69 kV)

The power flow from bus A to bus B is:

S_{AB} = $\sqrt{3}$ $V_A \cdot \hat{I}_{AB}$ (the caret
on $\hat{I}$ means *conjugate*)   (Eq 8)

= $(\sqrt{3} \cdot 69 \underline{/0°})$ (48.73 $\underline{/+21.33°}$)

= 5823 $\underline{/21.33°}$ kVA

= P_{AB} + j Q_{AB}

= 5425 + j2118

The current I_{AB} on the 13.5 kV side is that at the 69 kV side multiplied by the transformation ratio:

I_{AB} = 48.73 $\underline{/-21.33°}$ $\cdot \dfrac{69}{13.8}$

= 243.7 $\underline{/-21.33°}$ (at 13.8 kV)

Next find the power flow from bus B to bus A:

S_{BA} = $-\sqrt{3}$ $V_B \cdot \hat{I}_{AB}$ (at 13.8 kV)

= $- (\sqrt{3} \cdot 13.60 \underline{/-1.6°})$

$\cdot (243.7 \underline{/21.33°})$

= – 5741 $\underline{/19.73°}$ kVA

= $P + j Q$

= –5403 – j1938

The transformer losses can now be found by adding the power from A to B and that from B to A:

S_{Losses} = S_{AB} + S_{BA}

= (5425 + j2118)

+ (–5403 – j1938)

= 22 + j180

The losses are 22 kW and 180 kvar between buses A and B. Figure 17 summarizes these results.

This example illustrates that once the bus voltages are known, the remaining calculations are straightforward.

The real problem of analyzing even a modest size system is the determination at the bus voltages. The loads are known but mostly non linear and complicated further by the fact that the reference voltage is often many buses away from the loads. In order to find the bus voltages, one has to resort to a cut-and-try method. The computer is an effective tool for this method since it can complete the set of calculations shown above much faster than if done by hand.

4.5 Impedance Diagram. Several things remain to be said about Figs 15, 16, and 17.

(1) Obviously all three diagrams show a single-phase equivalent of the three-phase system. The conditions for this equivalence to be true were covered in Section 3.

(2) The impedance diagram of Fig 16 is the correct way to represent graphically resistances and reactances. However, it is felt that drawing the graphical sym-

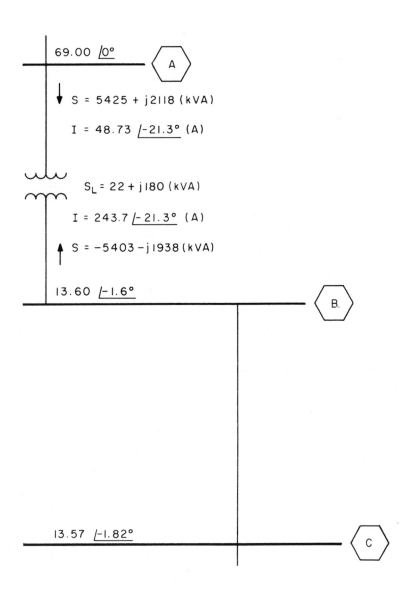

Fig 17
Flow Diagram

bols of resistances, inductances and capacitances is superfluous since the expressions for impedances alongside a straight line do describe the line elements per-

fectly well. Along this line, the graphical representation described by Fig 18 is suggested as a method for bridging the gap between the single line diagram and

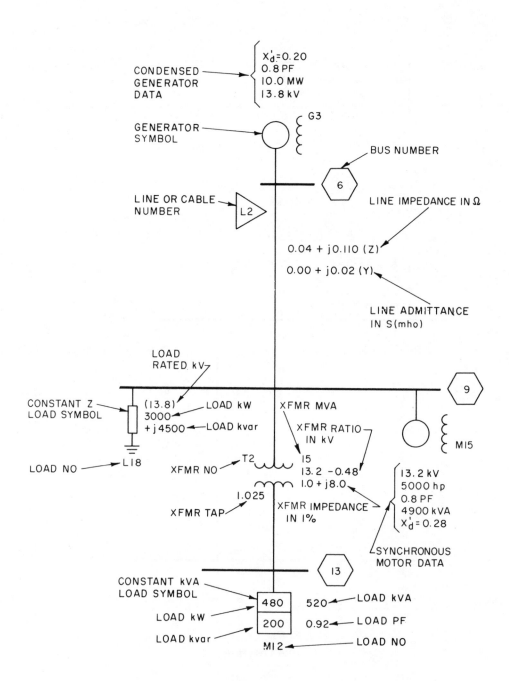

Fig 18
Suggested Format
Raw Data Diagram

the computer input document. A skeleton drawing of the power system showing buses, lines, generators, and loads, each with its assigned numbers, can be duplicated for multiple use as impedance diagrams and as a flow diagram. It should be noted that these diagrams are working tools and as such do not require standardization. However, the analyst should adopt a method suitable for keeping track of masses of data, for even small system studies require and generate a large amount of information.

(3) In power systems analysis the term *bus* does not always have the meaning understood by a plant electrician, for instance. The analyst calls *bus* any point of the system where voltages are calculated. The term is interchangeable with *node*. Fictitious *buses* may be introduced on the network to obtain voltage solutions at certain points of interest. An example of this may be a 150 mi transmission line broken down in 5 sections of 30 miles (that is, a *bus* introduced every $1/5$ of the length) in order to avoid the complicated but exact model of the long line.

(4) In the same vein, the term *line* is often given the more general meaning of *branch*, that is, any element between two nodes. For example, the transformer data will be entered on a computer input document called *line data*.

4.6 Extent of the Model

4.6.1 General. No rigid rules can be established on how much of a power system should be modeled for a given study. The system analyst has to exercise judgment and develop a feel for this as he gains experience.

The capability of available computer programs will not likely be the limiting factor in industrial and commercial power system applications. There are load flow programs that can handle up to 6000 buses. It is also probable that the engineer's time to prepare the data will cost more than the computer time.

A rough guideline would be: when in doubt, model the more extensive system. Of course, there are penalties in doing so:

(1) More data preparation time required
(2) Increased possibilities of input error
(3) More output data generated
(4) Significant results become a smaller section of the output

The objectives of the study should always be kept in focus. This will help in eliminating useless work.

4.6.2 Utility Supplied Systems. A large number of industrial and commercial establishments are supplied by stiff utility systems. Stiffness is a relative function of the size of the plant load and local generation. If the external power system or utility is large compared with that of the plant, disturbances within the plant do not affect the voltage at the point of connection. In such a case the utility system is said to be an infinite system. The connection point will be an infinite bus.

This concept can be extended within the plant electrical distribution system when studies are concerned with small areas electrically remote from the utility supply. Conversely sections of utility systems may require modeling in cases where this stiffness does not exist. It is, therefore, important that a sound knowledge of the utility supply systems be acquired before going ahead full bore with studies.

4.6.3 Isolated Systems. The question of whether an isolated system should be modeled in full or in part is easier to determine. These are usually relatively small and as such could be represented fully for most kinds of studies. The extra effort of gathering a set of data for the entire system, even though a smaller section would suffice, will not be lost since the additional data will be used in some future study. The nature of an isolated system is such that a modification or a disturbance is more apt to be felt throughout the system.

4.6.4 Swing Bus. The requirement that a reference voltage and angle be specified at one bus for a network solution to be possible (see 4.4) introduces the concept of *swing* or *slack* bus.

For the network to be in equilibrium at any instant the total generation must equal the total load plus the total losses. Since the *line* losses are not specified as an input, at least one bus of the network has to be capable of adjusting the generated power for the equilibrium to be achieved.

This bus is called the swing or slack bus. It is usually the bus assigned the fixed voltage and reference angle. It is usually a generator bus in the case of isolated systems, or the *infinite* bus behind the source impedance for a utility supplied system.

The system analyst should always specify the swing bus. It should be that bus in the system that maintains the closest voltage regulation.

4.7 Models of Branch Elements

4.7.1 Lines. Four parameters affect the performance of a line connecting a source to a load: series resistance, series inductance, shunt capacitance, and shunt conductance. A short length of conductor can be modeled as in Fig 19. A line can be considered as many short lengths of conductors placed in series to yield the model of Fig 20(a). The individual lengths of conductor could be made shorter thus increasing the number of lengths for a given length of line. Continuing this process to the limit defines the model called *line with distributed constants*. This model has been reduced to the equivalent circuit shown in Fig 20(b), where the series arm is defined by

$$Z' = Z \frac{\sinh \gamma \ell}{\gamma \ell} \qquad \text{(Eq 9)}$$

and the shunt branches by:

$$\frac{Y'}{2} = \frac{Y}{2} \frac{\tanh (\gamma \ell/2)}{(\gamma \ell/2)}$$
$$= \frac{1}{Z_C} \frac{(\cosh \gamma \ell - 1)}{\sinh \gamma \ell} \qquad \text{(Eq 10)}$$

Two figures of merit appear in these equations:

Fig 19
Equivalent Circuit of Short Conductor

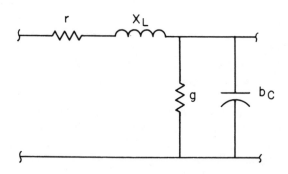

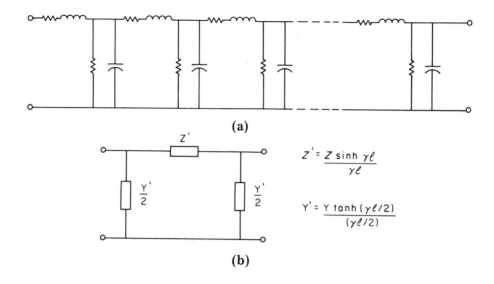

Fig 20
(a) Line with Distributed Constants
(b) Long Line Equivalent Circuit

$Z_C = \sqrt{z/y}$, defined as the *characteristic or surge impedance* of the line   (Eq 11)

$\gamma = \sqrt{y\,z}$, defined as the *propagation constant*   (Eq 12)

Both Z_C and γ are complex numbers. The propagation constant γ can be expressed in the rectangular form:
$\gamma = \alpha + j\beta$

This defines: (Eq 13)

α = attenuation constant
β = phase constant (in radians)

The other variables are:

l = total length of line
r = conductor effective resistance in ohms per unit of length
x = conductor series inductive reactance in ohms per unit of length

g = shunt conductance to neutral in siemens (mho) per unit of length
b = shunt capacitive susceptance in siemens (mho) per unit of length
$X = 2\pi fL$
$y = \tfrac{1}{2}\pi fC$
L = conductor total inductance in henry per unit of length
C = conductor shunt capacitance in farads per unit of length
$z = r + jx$ = series impedance in ohms per unit of length
$y = g + jb$ = shunt admittance to neutral in siemens (mhos) per unit of length
Z = zl = total series impedance of line in ohms
Y = yl = total shunt admittance of line to neutral in siemens (mhos) per unit of length.
sinh, cosh, tanh = hyperbolic functions

These functions of complex numbers

can be evaluated by using the following relationships:

$$\sinh(\gamma\ell) = \sinh(\alpha\ell+j\beta\ell)$$
$$= \sinh(\alpha\ell)\cos(\beta\ell) \quad \text{(Eq 14)}$$
$$+ j\cosh(\alpha 1)\sin(\beta 1)$$
$$\cosh(\gamma\ell) = \cosh(\alpha 1)\cos(\beta\ell)$$
$$+ \sinh(\alpha\ell)\sin(\beta\ell) \quad \text{(Eq 15)}$$

$$\tanh X = \frac{\sinh X}{\cosh X} \quad \text{(Eq 16)}$$

$$\sinh(\alpha\ell) = \frac{e^{\alpha\ell} - e^{-\alpha\ell}}{2} \quad \text{(Eq 17)}$$

$$\cosh(\alpha\ell) = \frac{e^{\alpha\ell} + e^{-\alpha\ell}}{2} \quad \text{(Eq 18)}$$

The surge impedance Z_C is approximately 400 Ω for single circuit power lines. The distributed constants model is valid for short or long lines at power or communication frequencies.

4.7.1.1 Long Lines. Power frequency overhead lines in excess of 150 mi could be represented by the distributed constants model reduced to an equivalent π as shown in Fig 20(b).

The shunt conductance g may be neglected since the dielectric is air (a good dielectric) and the conductor spacing is large. However, if the corona losses are important, they may be represented as a value for g. Computer programs will readily accept the data for Z' expressed in the rectangular form, that is, the equivalent series resistance R' and the equivalent series inductive reactance X'. It should be noted that even though g is neglected ($g = 0$), a non zero value of G' will appear in the equivalent circuit of Fig 20B. The values of $G'/2$ and $B'/2$ (in siemens) may have to be modified to MW or Mvar to suit the program input requirements. (The computer treats them as constant impedance loads.) If the computer requires a line *charging* *Mvar* the value B' and not $B'/2$) must be used to calculate:

$$\text{Mvar} = 3(kV)^2 \cdot B' \quad \text{(Eq 19)}$$

The program will internally assign half of that value to the bus at each end of the line.

The value of $G'/2$ can be modified to:

$$\text{MW} = 3(kV)^2 \cdot \frac{G'}{2} \quad \text{(Eq 20)}$$

and input as a constant impedance bus load.

The voltage (kV) is the line-to-line voltage corresponding to the base voltage used in the program input document.

4.7.1.2 Medium Lines. In the range of 50 to 150 mi approximately, little accuracy is lost by simplifying Eqs 9 and 10 to

$$Z' = Z$$
$$\frac{Y'}{2} = \frac{Y}{2}$$

Thus neglecting the products $\dfrac{\sinh\gamma\ell}{\gamma\ell}$ and $\dfrac{\tanh(\gamma\ell/2)}{(\gamma\ell/2)}$ yields the model of Fig 21(a), called the nominal π circuit. In this model the shunt branches are purely capacitive (no inductance).

The nominal π circuit can be thought of as being formed by the process described at the beginning of 4.7.1, with the exception that the unit length of conductor is made longer (instead of shorter), and the circuit made symmetrical (bilaterally). This results in the constants being lumped by an approximation process. The nominal T circuit is formed the same way, with the difference that all the shunt constants are lumped into one as compared with the π circuit where series constants are lumped into a single one.

Use of the nominal T model is not popular since it requires addition of a

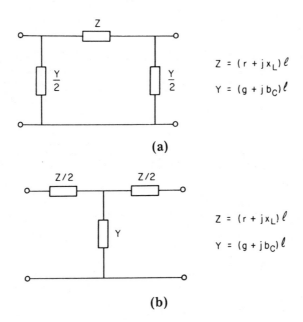

$$Z = (r + jx_L)\ell$$
$$Y = (g + jb_C)\ell$$

$$Z = (r + jx_L)\ell$$
$$Y = (g + jb_C)\ell$$

(a)

(b)

Fig 21
Medium Line Equivalent Circuits
(a) Nominal π (b) Nominal T

fictitious bus in the middle of the line. Entering data into the program, for the nominal π circuit follows the same requirements as for the long line model.

4.7.1.3 Short Lines. For overhead lines shorter than 50 mi, neglecting the shunt capacitance in the models presented earlier will not greatly affect the results of load flow, short-circuit or stability calculations. This yields the model of Fig 22.

4.7.2 Cables. The overhead line models are equally applicable to cables. While the resistances are substantially the same, the relative values of reactances are vastly different. Table 4 compares two cases, one at 69 kV, the other at 13.8 kV. The cable inductive reactance is about $1/4$ that of the line, but the capacitive reactance is 30 to 40 times that of the line.

This comparison suggests that the medium line model, the nominal π, should be used for cables in the order of one mile in length (approximately $1/40$ of 50 mi). The shorter the cable run the better the accuracy when using this model. However, the computer program used may have a limitation as to how small a quantity it will accept.

It is doubtful that any medium voltage system will have feeder lengths requiring representation of the capacitive reactance.

Fig 22
Short Line Equivalent Circuit

$$Z = (r + jx_L)\ell$$

Table 4
Comparison of Overhead Lines and Cable Constants

	Values in Ω/mile for 500 kcmil Cu Conductors			
	69 kV		13.8 kV	
	Overhead Line*	Cable**	Overhead Line†	Cable‡
Resistance	0.134	0.134	0.134	0.134
Inductive reactance	0.695	0.176	0.613	0.146
Capacitive reactance	0.162×10^6	0.005×10^6	0.142×10^6	0.003×10^6

  *Open-wire equilateral conductor spacing of 8 ft
 **Three-conductor oil-filled paper-insulated cable rated 69 kV
  †Open-wire equilateral conductor spacing of 4 ft
  ‡Three-conductor oil-filled paper-insulated cable rated 15 kV

Table 5
Conductor Data

Material	% Conductivity	Application
Copper		
soft	100	cable construction
soft, tinned	93.15 to 97.3	cable construction
hard, shape	98.4	bus bar
hard, round	97.0	overhead line conductors
Aluminum	61.0	cables, bars, tubes
Aluminum alloys		
5005-H19	53.5	overhead line conductors
6201-T81	52.5	overhead line conductors
Galvanized steel		
Siemens-Martin	12.0	
high strength	10.5	
extra high strength	9.4	

4.7.3 Determination of Constants. Electrical conductor characteristics are available from numerous sources and need not be repeated here. A few general comments are in order.

4.7.3.1 Resistance. The *effective resistance* of the conductors should be used. The effective resistance takes into account the conductor:

(1) Material
(2) Size
(3) Shape
(4) Temperature
(5) Frequency
(6) Environment

Copper and aluminum are the most used conductor materials for lines and cables. Soft annealed chemically pure copper has 100% conductivity (IACS standards). This is equivalent to 875.2 Ω for a one mile long round wire weighing one pound at 20 °C. All other materials can have their conductivities expressed as a percentage of the standard, a few of which are listed in Table 5.

The conductor resistance will vary with temperature according to the following formula:

$$R_2 = R_1\left[1 + \alpha(t_2 - t_1)\right] \qquad \text{(Eq 21)}$$

R_2 = resistance at temperature t_2
R_1 = resistance at temperature t_1
α = temperature coefficient per degree at temperature t_1.

At 20 °C, the coefficient α per degree Celsius is:
Copper: 0.00393
Aluminum: 0.00403
Galvanized steel:
    SM: 0.0039
    HS: 0.0035
    EHS: 0.0032

It is not possible to predict the exact operating conductor temperature if the conductor current is not known. The analyst has the choice of either estimating the conductor temperature or assuming the worst case, which, in some studies, might be the maximum allowable temperature of the cable. Other studies might require that the minimum conductor temperature be used for the worst case.

The ac resistance of conductors is higher than the dc resistance due to skin effects. The effect is more pronounced as the conductor cross section or the operating frequency increases. Conductor data tables usually include ac resistances at power frequencies. The skin effect is a major factor in the design of high current (several thousand amperes) ac bus systems, such as for electric furnaces.

The flux established by alternating current in a conductor may link other conductors or metallic masses in its proximity thus generating voltages in those parts. These voltages may cause currents to flow through closed circuits and thus cause I^2R losses other than those of the conductor itself. These losses can be represented as an additional component of resistance in series with the conductor resistance. The reader should consult [1] and [2] for information on this subject.

4.7.3.2 Inductive Reactance. The inductive reactance of a circuit has two components: that due to its own circuit (self) and that due to other circuits in its vicinity (mutual). The inductance of a conductor also has two components: that caused by the current in itself and that caused by the currents in other conductors of the same circuit. Finally the inductance of a conductor due to its own current is divided in two parts: the first part considers the flux internal to the conductor; the second part, the flux external to the conductor. This last division has been modified to simplify tables of conductor characteristics. Tables list the conductor inductive reactance X_s at one foot spacing even if the actual spacing is larger or smaller than one foot.

A second table, valid for any type or size of conductor, lists spacing factors X_d which, added to the one foot reactance will give the correct total reactance for the given circuit conductor spacing. The spacing factor table is calculated from the equation:

$$\begin{aligned} x_d &= 4.657 \cdot 10^{-3} \cdot f \cdot \log \text{GMD} \\ &= \Omega/(\text{conductor} \cdot \text{mile}) \qquad \text{(Eq 22)} \end{aligned}$$

where

GMD = geometric mean distance of the conductors
    For three conductors spaced d_1, d_2, d_3

$$\text{GMD} = \sqrt{d_1 \cdot d_2 \cdot d_3}.$$

Note that in Eq 22 a GMD smaller than 1 yields a negative spacing factor.

Cables in steel conduit exhibit higher reactances than in free air. The calculations are too complex to develop by hand, hence, the curves presented in Section 4 of [1] and Section .691 of [21] should be used for estimating purposes. Tables are also available in Chapter 1 of [17].

4.7.3.3 Shunt Capacitive Reactance. The determination of the capacitive reactance follows the same pattern as the inductive reactance. Conductor tables give the value of reactance X' at one-foot spacing. A spacing factor X' is added to X' to yield the total capacitive reactance of the conductor. Spacing factor tables are calculated from:

$$x'_d = \frac{4.099}{f} \cdot 10^{-6} \log \text{GMD}$$

$$= \Omega/(\text{conductor} \cdot \text{mile}) \qquad (\text{Eq 23})$$

The capacitive reactance of shielded cables is determined from:

$$x'_c = \frac{1.79 \, G \cdot 10^6}{f \cdot k}$$

$$= \Omega/(\text{phase} \cdot \text{mile}) \qquad (\text{Eq 24})$$

where

G = geometric factor
k = dielectric constant of cable insulation
f = frequency

$$G = 2.303 \log \frac{2r}{d} \qquad (\text{Eq 25})$$

where

r = inside diameter of shield
d = outside diameter of conductor

Typical values of k are 6.0 for rubber, 5.0 for varnished cambric, 2.6 for polyethylene, and 3.7 for paper.

4.7.4 Reactors. Reactors are used as branch elements in the following applications:

(1) To limit current during fault conditions

(2) To buffer cyclic voltage fluctuations caused by repetitive loads (in conjunction with condensers)

(3) To limit motor starting currents

They are modeled as impedances consisting of an inductive reactance in series with a resistance expressed as $R + jX$. Manufacturers' design or test data should be obtained for existing applications.

The resistance section of this model can usually be neglected in motor starting studies since it is small with respect to the reactance and the power factor of the motor is low during starting.

4.7.5 Capacitors. Series capacitors are sometimes used on transmission and distribution lines to compensate for the inductive reactance drop or to improve the system stability by increasing the amount of power that can be transmitted on tie lines. They are represented by a negative reactance of the form $0 - jX$, in series with the line impedance.

For capacitors specified in microfarads per phase, the reactance may be expressed in the general form:

$$X = \frac{10^6}{2 \pi fc} = \Omega/(\text{per phase}) \qquad (\text{Eq 26})$$

When specified in kilovars per phase (Q_C), the capacitor voltage rating (V_C) must also be known to calculate:

$$X = \frac{V_C}{Q_C} \cdot 10^3 = \Omega/(\text{per phase}) \qquad (\text{Eq 27})$$

It should be noted that the series capacitor voltage rating is a function of the amount of compensation of the design and will generally be a fraction of the system line-to-neutral voltage. The application of series capacitors should always be accompanied by thorough studies, since it is easy to create destructive overvoltage and ferro-resonance conditions.

4.7.6 Transformers

4.7.6.1 Two-Winding Transformers. The equivalent circuit of a transformer is shown in Fig 23(a). The rectangle represents an ideal voltage transformation ratio $n_s/n_p = N$, where n_s and n_p are the number of turns of the primary and secondary windings respectively. R_p and R_s are the effective resistances of the windings X_p, and X_s, their leakage reactances. G_0, the shunt conductance, models the iron losses that remain constant when the transformer is energized at rated voltage and B_M, the shunt inductive susceptance, is equivalent to the quadrature magnetizing current at no load.

It can be demonstrated that Fig 23(a) is equivalent to Fig 23(b). In the latter the secondary resistance and reactance has been *reflected* to the primary side of the ideal transformer by multiplication with the inverse of the square of the turns ratio N. A close approximation of this circuit is possible by moving the shunt branch and combining the primary and secondary impedances as shown on Fig 24(a). Another simplification consists in eliminating the shunt branch altogether to yield Fig 24(b). In many types of studies, the resistance R_T, being small with respect to X_T, is also neglected thus reducing the model of the transformer to a single series reactance.

Transformer nameplate specifies the impedance Z_T and the transformation ratio. An assumption may be made that $X_T \cong Z_T$ and the single series reactance model used.

Fig 23
Two-Winding Transformer
Equivalent Circuits

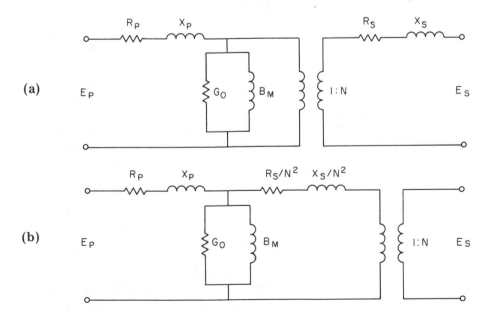

(a)

(b)

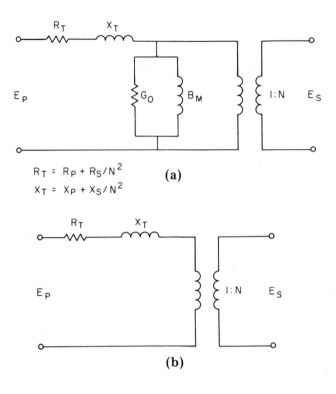

$$R_T = R_P + R_S/N^2$$

(a)

$$X_T = X_P + X_S/N^2$$

(b)

Fig 24
Two-Winding Transformer
Approximate Equivalent Circuits

Use of Fig 24(b) model requires that an estimate of R be made from typical data [1], and a value for X_T calculated from

$$X_T = \sqrt{Z^2 - R^2}$$

Transformer test data will usually be sufficient to calculate all the parameters for the circuits of Figs 23(a), 23(b) and 24(a). When maximum accuracy is needed the effective resistance R_T should include the winding resistances corrected for the operating temperature and another series resistance to account for stray losses [2], [3]. The model of Fig 23 necessitates the creation of a fictitious bus for entry of the shunt admittance data in the program.

4.7.6.2 Transformer Taps. Thus far, only single ratio transformers have been dealt with. In real life, transformers have taps, normally on the high voltage windings, to provide a voltage ratio best suited to the power system. The taps may be changeable automatically under load (LTC transformer) or fixed (manually changed in de-energized condition).

The resistance and leakage reactance of the tapped windings are slightly different at different taps. This may be ignored if the correct values are not known. On the other hand, transformer test data may specify values for the taps, in which case these values should be used. The main effect of changing taps is the change of voltage ratio and therefore the change of

(a)

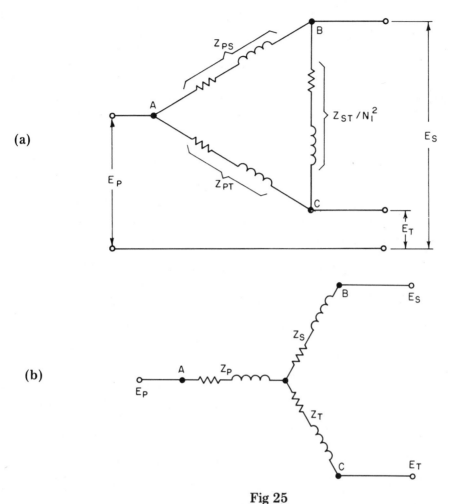

(b)

Fig 25
Three-Winding Transformer Approximate Equivalent Circuits
(a) Simplified-Delta   (b) Simplified-Wye

voltage base for which the impedance diagram should be prepared. This will be described in more detail in 4.8.

The analyst should pay particular attention to the specific requirements of programs for specifying taps. For instance the tap value 1.05 per unit (105%), interpreted as an additional 5% to the voltage ratio, yields opposite results if applied to the opposite sides of the transformer.

Once the data is input for a given condition and a certain voltage on a specified

bus is required, the computer program will, upon request, automatically adjust the taps and modify the system impedances as necessary for the new turns ratio.

4.7.6.3 Three-winding Transformers. The circuits of Figs 23 and 24 apply to any two windings of the three-winding transformers.

The three possible combinations, put together using Fig 24(b) as basic model, form a delta as shown on Fig 25(a), where the new subscript T denotes the tertiary winding. Elimination of the ideal transformer symbols is made possible by calling the voltages at nodes A, B and C, 1.0 per unit and remembering that those nodes have base voltage values E_p, E_s and E_t respectively. Assuming that the 3-phase system is balanced, the fictitious neutral will be common to all three voltage levels and can be eliminated to simplify the diagram.

The loop formed by the delta (Δ) circuit can also be eliminated. This requires making a delta-wye transformation. The end result shown in Fig 25(b) is a format acceptable to computer programs. The new node D at the center of the Y creates a fictitious bus that cannot be identified physically but that is necessary to identify the impedance values to the computer.

The relationships for reduction from Δ to Y are:

$$Z'_p \doteq \tfrac{1}{2}(Z_{ps} + Z_{pt} - \frac{1}{N_1^2} Z_{st}) \qquad \text{(Eq 28)}$$

$$Z'_s = \tfrac{1}{2}(\frac{1}{N_1^2} Z_{st} + Z_{ps} - Z_{pt}) \qquad \text{(Eq 29)}$$

$$Z'_t = \tfrac{1}{2}(Z_{pt} + \frac{1}{N_1^2} Z_{st} - Z_{ps}) \qquad \text{(Eq 30)}$$

The relationship as a function of the Y values is:

$$Z_{ps} = Z'_p + Z'_s \qquad \text{(Eq 31)}$$

$$Z_{pt} = Z'_p + Z'_t \qquad \text{(Eq 32)}$$

$$Z_{st} = N_1^2 (Z'_s + Z'_t) \qquad \text{(Eq 33)}$$

4.7.6.4 Phase-Shifting Transformers. Consider again the example of 4.4. Assume this time that the voltage at bus B

has the same magnitude but that the angle is shifted back another $2°$ by the winding arrangement within the transformer, and that the rest of the information on Fig 17 is the same as before. The new impedance diagram will appear as shown in Fig 26(a).

Now resolve as before with $V_B = 13.60$ $\underline{/-1.6°} -2.0°$. The results are shown in Fig 26B. Note that the phase shift of a mere $2°$ has caused the real power from A to B to jump from 5425 to 22 803 kW, but the reactive power decreased from 2118 to 1605 kvar.

This example illustrates the main purpose of phase-shifting transformers to control the flow of real power between two buses.

These transformers usually have load tap changing mechanisms that will vary the phase angle between primary and secondary automatically or manually. Thus computer programs will be set up to vary the amount of angle shift (plus or minus), within limits specified in the input, to achieve the desired amount and the direction of real power flow.

The data required by computer programs will generally include the following:

(1) Center ($0°$ shift) position impedance
(2) Positive limit position impedance
(3) Negative limit position impedance
(4) Angle shift interval between taps
(5) Number of taps

Program subroutines allow the computer to automatically estimate the value of the impedances at the intermediate taps.

4.7.6.5 Other Transformer Models. The foregoing discussion has been intended to give the reader a running start on the subject of transformer modeling. It is far from being complete.

4.8 Power System Data Development
4.8.1 Per Unit Representations. Net-

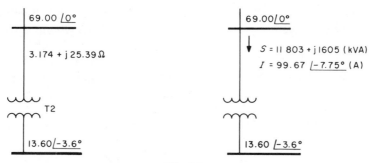

Fig 26
Impedance and Flow Diagrams
(a) Impedance Diagram   (b) Flow Diagram

work calculations can be made with impedances, voltages and currents expressed either in actual ohms, volts or amperes or in the dimensionless per unit method. While both representations will yield identical results, the per unit method is generally preferred as it will do the job much more conveniently especially if the system being studied has several different voltage levels. Also, the impedances of electric apparatus are usually given in per unit or percent by manufacturers.

The per unit value of any quantity is its ratio to the chosen base quantity of the same dimensions, expressed as a dimensionless number. For example, if base voltage is taken as 4.16 kV, voltages of 3740 V, 4160 V, and 4330 V will be 0.9, 1.00, and 1.04, respectively, when expressed in per unit on the given base voltage. The chosen base voltage, 4.16 kV, is referred to as base voltage, 100% voltage or unit voltage.

There are four base quantities in the per unit system: base kVA, base volts, base ohms, and base amperes. They are so related that the selection of any two of them determines the base values of the remaining two. It is a common practice to assign study base values to kVA and voltage. Base amperes and base ohms

are then derived for each of the voltage levels in the system.

In power system studies, the base voltage is usually selected as the nominal system voltage at one point of the system such as the voltage rating of a generator or the nominal voltage at the delivery point of the utility supply. The base kVA is similarly taken as the kVA rating of one of the predominant pieces of system equipment such as a generator or a transformer, but usual practice is to choose a convenient round number such as 10 000 for the base kVA. The latter selection has some advantage of commonality when several studies are made, while the former choice means that at least one significant component will not have to be converted to a new base.

The basic relationships for the electrical per unit quantities are as follows:

$$\text{Per unit volts} = \frac{\text{actual volts}}{\text{base volts}} \qquad \text{(Eq 34)}$$

$$\text{Per unit amperes} = \frac{\text{actual amperes}}{\text{base amperes}}$$

$$\text{(Eq 35)}$$

$$\text{Per unit ohms} = \frac{\text{actual ohms}}{\text{base ohms}} \qquad \text{(Eq 36)}$$

For three-phase systems, the nominal line-to-line system voltages are normally used as the base voltages. The base kVA is assigned the three-phase kVA value. The derived values of the remaining two quantities are:

$$\text{Base amperes} = \frac{\text{base kVA}}{\sqrt{3}\ \text{base kV}} \qquad (\text{Eq } 37)$$

$$\text{Base ohms} = \frac{(\text{base kV})^2}{\text{base MVA}} \qquad (\text{Eq } 38)$$

It is convenient in practice to convert directly from ohms to per unit ohms without first determining base ohms according to the following expression:

$$\text{Per unit ohms} = \frac{\text{ohms} \cdot \text{base MVA}}{(\text{base kV})^2}$$

$$(\text{Eq } 39)$$

For a three-phase system, the impedance is in ohms to neutral and the base kVA is the three-phase value.

Where two or more systems with different voltage levels are interconnected through transformers, the kVA base is common for all systems, but the base voltage of each system is forced by the turns ratio of the transformer connecting the systems, starting from the one point for which the base voltage has been declared. Base ohms and base amperes will thus be correspondingly different for systems of different voltage levels.

Once the system quantities are expressed as per unit values, the various systems with different voltage levels can be treated as a single system and the necessary variables can be solved. Only when reconverting the per unit values to actual voltage and current values is it necessary to recall that different base voltages exist throughout the system.

When impedance values of devices are expressed in terms of their own kVA and voltage ratings which differ from the base values of a circuit, it is necessary to refer these values to the system base values. This may also happen when machines rated at one voltage may actually be used in a circuit at a different voltage. In such cases, the per unit impedance of the device must be changed to either a new base kVA or new base voltage, or both, by the equation:

$$\text{Per unit } Z_2 = \text{per unit } Z_1$$
$$\cdot \frac{(\text{base kV}_1)^2}{(\text{base kV}_2)^2} \frac{\text{base kVA}_2}{\text{base kVA}_1}$$

$$(\text{Eq } 40)$$

where subscripts 1 and 2 refer to the old and new base conditions, respectively.

4.8.2 Applications Example. A section of the power system described in 4.5 has been repeated in Fig 27 as an illustration of the per unit system. The transformer ratios were changed slightly to improve this example. The steps in reducing the data to per unit are as follows:

(1) Select base power: $S = 10\ 000$ kVA
(2) Determine base voltages
    (a) Select bus 2 nominal voltage of 69 kV as base
    (b) Calculate base voltages at other system levels

$$\text{Bus 4:}\quad \text{kV} = 69.0 \cdot \frac{13.8}{66}$$
$$= 14.427 \text{ kV}$$

$$\text{Bus 36:}\quad \text{kV} = 14.427 \cdot \frac{2.4}{13.2}$$
$$= 2.623 \text{ kV}$$

$$\text{Bus 37:}\quad \text{kV} = 14.427 \cdot \frac{0.48}{13.2}$$
$$= 0.525 \text{ kV}$$

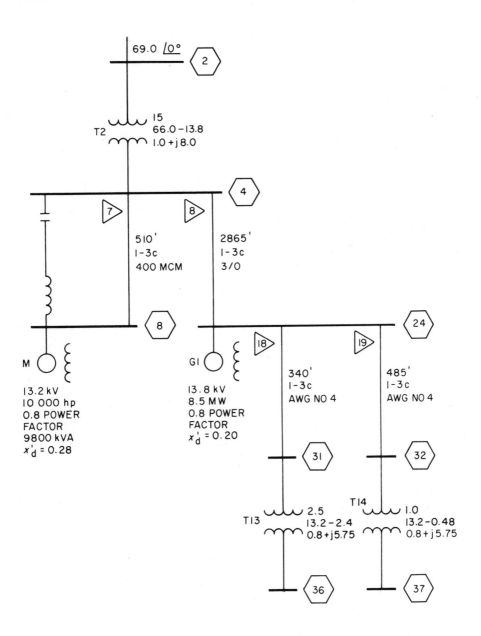

Fig 27
Impedance Diagram Raw Data

(3) Calculate base impedances using Eq 38

    (a) 69 kV system:

$$Z = \frac{69^2 \cdot 10^3}{10\ 000}$$

$$= 476.1\ \Omega$$

    (b) 13.8 kV system:

$$Z = \frac{14.427^2 \cdot 10^3}{10\ 000}$$

$$= 20.82\ \Omega$$

    (c) 2.4 kV system:

$$Z = \frac{2.623^2 \cdot 10^3}{10\ 000}$$

$$= 0.688\ \Omega$$

    (d) 0.48 kV system:

$$Z = \frac{0.525^2 \cdot 10^3}{10\ 000}$$

$$= 0.02756\ \Omega$$

(4) Calculate base currents using Eq 37

    (a) 69 kV system:

$$I = \frac{10\ 000}{\sqrt{3} \cdot 69.0}$$

$$= 83.67\ A$$

    (b) 13.8 kV system:

$$I = \frac{10\ 000}{\sqrt{3} \cdot 14.427}$$

$$= 400.2\ A$$

    (c) 2.4 kV system:

$$I = \frac{10\ 000}{\sqrt{3} \cdot 2.623}$$

$$= 2201\ A$$

    (d) 0.48 kV system:

$$I = \frac{10\ 000}{\sqrt{3} \cdot 0.525}$$

$$= 11\ 000\ A$$

(5) Summarize the base data in **Table 6**

(6) Convert transformer impedances to per unit using Eq 40

    (a) T2:

$$Z = \frac{1.0 + j8.0}{100} \cdot \frac{66^2}{69^2} \cdot \frac{10}{15}$$

$$= 0.006\ 10 + j0.048\ 80$$

or

$$Z = \frac{1.0 + j8.0}{100} \cdot \frac{13.8^2}{14.427^2} \cdot \frac{10}{15}$$

$$= 0.006\ 10 + j0.048\ 80$$

    (b) T13:

$$Z = \frac{0.8 + j5.75}{100} \cdot \frac{13.2^2}{14.427^2} \cdot \frac{10}{2.5}$$

$$= 0.026\ 79 + j0.19254$$

or

$$Z = \frac{0.8 + j5.75}{100} \cdot \frac{2.4^2}{2.623^2} \cdot \frac{10}{2.5}$$

$$= 0.026\ 79 + j0.192\ 55$$

    (c) T14:

$$Z = \frac{0.8 + j5.75}{100} \cdot \frac{13.2^2}{14.427^2} \cdot \frac{10}{1}$$

$$= 0.066\ 97 + j0.481\ 35$$

or

Table 6
System Base Values
(Base Power 10 000 kVA)

Bus	Base kV	Base Z	Base I
2	69.00	476.1	83.67
4	14.427	20.82	400.2
8	14.427	20.82	400.2
24	14.427	20.82	400.2
31	14.427	20.82	400.2
32	14.427	20.82	400.2
36	2.623	0.688	2201.0
37	0.525	0.027 56	11 000.0

Table 7
Cable Data

Line	Length (ft)	Conductor Size	r (Ω/1000') (ft)	x_1 (Ω/1000') (ft)	Conductor Outside Diameter (inches)	Insulation Thickness (inches)
7	510	400 MCM	0.0297	0.0370	0.728	0.175
8	2865	3/0	0.0668	0.0423	0.470	0.185
18	340	4	0.2992	0.0516	0.232	0.220
19	485	4	0.2992	0.0516	0.232	0.220

$$Z = \frac{0.8 + j5.75}{100} \cdot \frac{0.48^2}{0.525^2} \cdot \frac{10}{1}$$

$$= 0.066\ 87 + j0.480\ 65$$

(7) Calculate line impedance in ohms

(a) Lines 7, 8, 18 and 19 are 3/c, copper cables, paper insulated, shielded conductors; dielectric constant: 3.7. See Table 7.

(b) Line 7:

$R = 0.0297 \cdot 510/1000$

$\quad = 0.015\ 15\ \Omega$

$X_L = 0.0370 \cdot 510/1000$

$\quad = 0.018\ 87\ \Omega$

$X_C =$ (neglect due to short length)

(c) Line 8:

$R = 0.0668 \cdot 2865/1000$

$\quad = 0.191\ 38\ \Omega$

$X_L = 0.0423 \cdot 2865/1000$

$\quad = 0.121\ 19\ \Omega$

From equations 24 and 25

$$X_C = \frac{1.79 \cdot 10^6}{60 \cdot 3.7} \cdot 2.303 \log$$

$$\cdot \frac{(0.470 + 2 \cdot 0.185)}{0.470}$$

$$= 4683\ \Omega/\text{mi}$$

(d) Line 18:

$X_C = 4683 \cdot 5280/2865$

$\quad = 8630\ \Omega$

$R = 0.2992 \cdot 340/1000$

$\quad = 0.101\ 73\ \Omega$

$X_L = 0.0516 \cdot 340/1000$

$\quad = 0.017\ 54\ \Omega$

$X_C =$ (neglect due to short length)

(e) Line 19:

$R = 0.2992 \cdot 485/1000$

$\quad = 0.145\ 11\ \Omega$

$X_L = 0.0516 \cdot 485/1000$

$\quad = 0.025\ 03\ \Omega$

$X_C =$ (neglect due to short length)

(8) Calculate line impedances in per unit with Eq 36

(a) Line 7:

$$Z = \frac{0.015\ 15 + j0.018\ 87}{20.82}$$

$$= 0.000\ 728 + j0.000\ 906\ \text{per unit}$$

(b) Line 8:

$$Z = \frac{0.191\ 38 + j0.121\ 19}{20.82}$$

$$= 0.009\ 192 + j0.005\ 821\ \text{per unit}$$

$$Y = \frac{Z_C}{-jX_C} = \frac{20.82}{-j8630}$$

$$= 0 + j0.002\ 413 \text{ per unit}$$

(c) Line 18:

$$Z = \frac{0.101\ 73 + j0.017\ 54}{20.82}$$

$$= 0.004\ 886$$
$$+ j0.000\ 842 \text{ per unit}$$

(d) Line 19:

$$Z = \frac{0.145\ 11 + j0.025\ 03}{20.82}$$

$$= 0.006\ 97$$
$$+ j0.001\ 202 \text{ per unit}$$

(9) Calculate X_d' of two synchronous machines in per unit with Eq 40

(a) Synchronous motor on bus 8

$$X_d' = j0.28 \cdot \frac{13.2^2}{14.427^2} \cdot \frac{10\ 000}{9800}$$

$$= j0.2392$$

(b) Generator $G1$

$$X_d' = j0.20 \cdot \frac{13.8^2}{14.427^2} \cdot \frac{10\ 000}{8500/0.8}$$

$$= j0.1722$$

The per unit data and the base voltages have been transferred to the impedance diagram of Fig 28, in readiness for the preparation of computer input document and as a record of the basic information for the study.

4.9 Models of Bus Elements

4.9.1 Loads in General. Power system loads may be classified by one or a combination of the following types, to account for their voltage dependence:

Constant power S
Constant impedance Z
Constant current I

Figure 29 shows their respective power/voltage relationships. A single expression:

$$\left(\frac{S}{S_i}\right) = \left(\frac{V}{V_i}\right)^k \quad \text{(Eq 41)}$$

can represent the three load types by making $k = 0$ for constant power, $k = 1$ for constant current and $k = 2$ for constant impedance loads. S_i is the initial power at V_i, the initial voltage, S the power at voltage V.

A more general expression can be formulated by expanding Eq 41 for the real and reactive power:

$$P + jQ = P_i \left(\frac{V}{V}\right)^{k_1} + jQ_i \left(\frac{V}{V}\right)^{k_2} \quad \text{(Eq 42)}$$

The subscript $_i$ has the same meaning as above. The exponents k_1 and k_2 could be different and nonintegers.

A load or group of loads could also be expressed in a more restrictive way by:

$$P + jQ = \left[A + B\frac{V}{V_i} + C\left(\frac{V}{V_i}\right)^2\right] P_i + j$$

$$\cdot \left[D + E\frac{V}{V_i} + F\left(\frac{V}{V_i}\right)^2\right] Q_i \quad \text{(Eq 43)}$$

again to reflect voltage dependency. In this case A, B, C and D, E, F represent fractions of P and Q respectively. The sums $A + B + C$ and $D + E + F$ must equal 1.

In stability studies, frequency, like voltage, may become an important factor in the modeling of loads. Linear frequency dependance would take the form

$$P + jQ = (1 + G\Delta f)P_i + j(1 + H\Delta f)Q_i \quad \text{(Eq 44)}$$

where G and H are the fractions of P_i and Q_i, respectively, being affected by the frequency deviation f from the steady state frequency.

The problem of assigning correct values to the constants (k_1, k_2, A through H) is

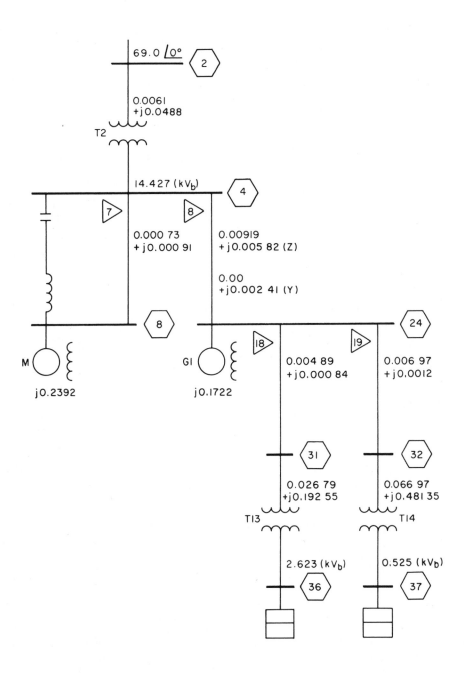

**Fig 28
Impedance Diagram Per Unit Data**

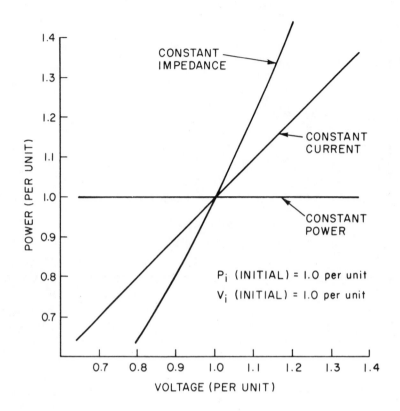

Fig 29
Effect of Voltage Variations for Three Types of Loads

very difficult when studying utility type systems, because the nature of the load is not known accurately. Additionally, the tasks of simulating the load in all its details would require a computer program of such size and cost that the effort would be prohibitive [4], [5], [6], [7].

Industrial and commerical power sys-tems are relatively modest in size. More-over bus loads are often arranged in such a way that grouping by type is easy to do, thus facilitating the preparation of computer input data and offering the possibility of combining large sections of the system to reduce the overall size of the study sections.

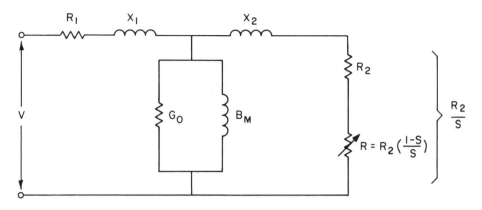

Fig 30
Induction Motor-Equivalent Circuit

4.9.2 Induction Motors. The induction motor single phase equivalent circuit is shown in Fig 30 [2], [3], [8]. The solution of this circuit for values of slip s at a given input voltage and frequency will yield, amongst other performance characteristics, three sets of curves, torque, current and power factor, typified in Figs 31, 32, 33.

If the three curves are known, the shaft output power can be calculated for any speed using the slip and torque with the equation:

$$\tau = 7.05 \cdot \frac{P_o}{(1 - S) N_s} \qquad \text{(Eq 45)}$$

where

τ = torque in lb-ft
P_o = output power in watts
N_s = synchronous speed in r/min

Input power can also be calculated using the current and power factor curves for the desired voltage.

$$S = 3VI (\cos \theta + j \sin \theta) \qquad \text{(Eq 46)}$$

$$\theta = \arccos PF \qquad \text{(Eq 47)}$$

where V is the line-to-neutral voltage.

In some cases the equivalent circuit constants will be known along with the input or output power. The analytical solution consists of finding the correct value of slip that will match the specified power. A cut-and-try method may be used here, but this will prove to be a tedious process if resolved by hand because of the repeated reduction of the complex network for each trial of slip value. Computers excel at cut-and-try methods, and for this reason programs will generally be written to accept this kind of input. The *equivalent circuit* model finds its major applications in motor starting and stability studies. At least one computer load flow program has been written to accept motor equivalent circuit data.

The curves of Figs 31, 32, 33 are used also in motor starting studies. The computer input data will consist of sets of values of torque, current and power factor taken at different intervals of slip along these curves. The mechanical load characteristics in the form of a torque–speed curve (in the same format as for the motor) will normally be required as well by the program. The motor and

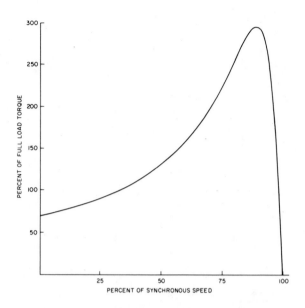

Fig 31
Induction Motor Torque Versus Speed

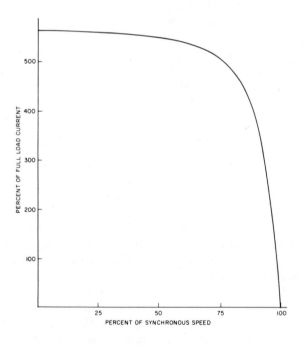

Fig 32
Induction Motor Current Versus Speed

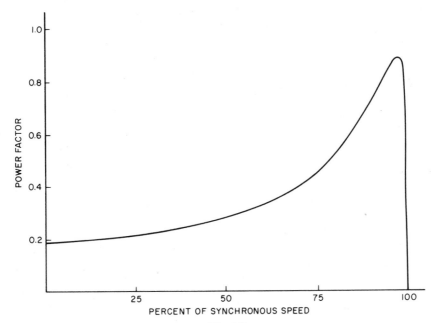

Fig 33
Induction Motor Power Factor Versus Speed

load moments of inertia will complement the description of the mechanical system.

The equivalent circuit constants can be calculated from motor test data or obtained from the manufacturer (calculated or test values). References [2], [3] and [9] describe the methods used for the calculations from test data.

4.9.2.1 Constant kVA Model. It may be appropriate in some study to represent a motor as a constant kVA load:

$$S = P + jQ = \text{constant}$$

This requires that the shaft horsepower (BHP), nameplate voltage, motor efficiency η, and power factor be known or estimated and substituted in the following equations:

$$P \text{ (kW)} = 746 \cdot \frac{\text{BHP}}{\eta} \qquad \text{(Eq 48)}$$

$$\theta = \arccos \text{PF} \qquad \text{(Eq 49)}$$

$$Q \text{ (kvar)} = P \tan \theta \qquad \text{(Eq 50)}$$

Both the efficiency and power factor are functions of motor voltage and percent load. An example is shown in Fig 34. Consequently if the objectives of the study warrant, it may be necessary to estimate the voltage in order to determine appropriate values for η and PF. Manufacturers publish motor characteristics (current, efficiency, power factor) at various loadings for a wide spectrum of motor sizes and types.

This model is generally applicable to load flow studies or other studies where the effect of model simplification is of secondary importance.

It should be emphasized that the reactive power Q is positive for an induction motor, even if the motor is operating in the generating mode, that is, negative slip.

77

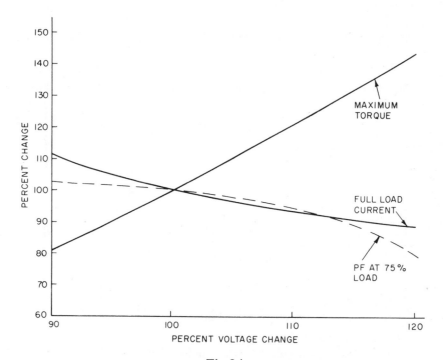

Fig 34
Effect of Voltage Variations on
Typical Induction Motor Characteristics

4.9.2.2 Models for Short-Circuit Studies. In this area of power system analysis standards specify the model to be used for motors to account for their contribution to short-circuit currents [9], [10]. The model is invariably a voltage source in series with an impedance as in Fig 35.

The voltage source is justified by the fact that at the very instant of a short circuit, the flux that exists at the air gap cannot change instantly. The energy that this flux represents has to dissipate itself in the form of current through resistance. This current flows in the electrical circuits linked by that flux, and is limited by the inductance of these circuits. The stator winding, being linked by that flux and having inductance, will therefore act as a limited source of current to the network connected to it. The flux will decay

rapidly, there being no source of current to maintain it. Thus, the time constant will be in the order of a few cycles only.

Fig 35
Model of Induction Motor
for Short-Circuit Study

$$R + jX = \frac{V_t}{\sqrt{3}\, I_{LR}}$$

V_t

The value of impedance in series with the voltage source is nearly the same value as the impedance which limits the locked rotor current when voltage is applied to the motor at rest. Accordingly, the impedance in Fig 35 can be calculated from:

$$R + jX = \frac{V}{\sqrt{3}\, I_{\mathrm{LR}}} \qquad (\text{Eq } 51)$$

where V is the motor nameplate line voltage and I_{LR} is the motor locked rotor current.

The resistance R is usually small relative to the reactance X. Moreover short-circuit currents have a low power factor. These two factors justify neglecting R in short circuit studies, where only the current magnitude is required.

When the short circuit study objective is to check circuit breaker capabilities, it is necessary to consider different conditions such as first cycle duty, momentary duty, interrupting duty, and breaker speed, in order to determine an appropriate value of motor series impedance that will account for the decay in current it contributes to the fault. This is spelled out in [22] and [23] and takes the form of impedance multiplying factors that vary as a function of time from the moment the fault occurred. The rationale for neglecting small motors' contributions is based on their remoteness, electrically, from the power circuit breakers and their very short time constants.

4.9.2.3 Constant Impedance Model. It is sometimes desirable to determine the system voltages at the instant a motor at rest is energized. The appropriate model is the equivalent circuit given earlier with the slip s equal to 1. This circuit is a constant impedance. It is generally simplified to a single impedance to ground and calculated from Eq 51.

4.9.3 Synchronous Machines. Synchronous machine models vary tremendously whether a study considers steady state or transient conditions. In this section the text will include both the simpler model for steady state conditions, and the more complicated ones for transient conditions.

4.9.3.1 Steady State Models

4.9.3.1.1 Generators. Once a power system has settled down after a change of any kind, generator output voltages will have been automatically brought back to the desired output values by the voltage regulator. The prime mover governor will also have taken a steady position to maintain the generator speed, consistent with the rest of the system, and to supply the amount of power programmed by its setting.

The generator output voltage will be a function of the field current. Its output power will be a function of the mechanical power applied to its shaft which will also be a function of the rotor angle that the field poles maintain relative to the revolving magnetic field in the stator.

Thus in a load flow study for steady state conditions, the generator can be modeled as a constant voltage source and a scheduled amount of kW. The system operating conditions may demand that a generator voltage output be adjusted automatically so that it supplies an amount of reactive power such that a bus, somewhere on the system, maintains a specified voltage. In this situation, the analyst specifies a range of reactive power (kvar) within which he is happy to operate the machine. The generator voltage will be adjusted by the program.

Other possibilities are:

(1) Specify a voltage and let the kW and kvar fall where they may. At least one bus of the system called the *swing*

bus will have to be specified for the specific voltage.

(2) Specify fixed kW and kvar and let the voltage fall where it may.

The last alternative suggests that a generator may be considered analytically as a negative constant power load. Many computer programs will accept negative power signs. Thus negative kW input to the bus load data would model a generator.

4.9.3.1.2 Synchronous Condenser. One difference between this machine and a generator is that the condenser will be a fixed load with very small kW value to represent its losses. Otherwise, it will generally be equipped with a voltage regulator similar to a generator and have its reactive power output, specified within certain limits, adjusted by the computer program to maintain a specified voltage at its own terminals or elsewhere in the system.

4.9.3.1.3 Synchronous Motors. The synchronous motors may or may not be equipped with regulators to control the excitation. Those equipped with regulators may control voltage, power factor, reactive power or even current, at their terminals or elsewhere. The real power in kW will be a function of the load driven by the motor, and will not be adjustable for the given set of conditions under study.

The analyst may, therefore, resort to modeling the motors as negative generators with reactive power limits, if a voltage or current control device is supplied. In the case of power factor or reactive power regulators, the specified kW will define a value of kvar using Eqs 52 through 54.

For motors with fixed excitation and a given fixed load, vee-curves must be used to calculate the equivalent reactive power. Figure 36 shows the V curve for a particular motor. The reactive power is calculated as follows:

(1) Draw a vertical line for the fixed excitation current

(2) Read the armature current I, at the intersection of the load line representing the motor running load, and the I line

(3) Estimate the motor terminal voltage (line to line)

(4) Calculate the reactive power

$$\text{kvar} = \sqrt{3(VI)^2 - \text{kW}^2} \qquad \text{(Eq 52)}$$

Alternately, if the power factor curves are shown on the graph, replace above steps 3 and 4 with:

(3) Read the power factor at the intersection of the load line and I line

(4) Calculate the reactive power

$$\text{kvar} = \text{kW} \tan \theta \qquad \text{(Eq 53)}$$

$$\theta = \arccos \text{PF} \qquad \text{(Eq 54)}$$

4.9.3.2 Short Circuit Models. The current contributed to a fault by a synchronous machine varies with time, from a high initial value to a moderate final steady state value. Equations 55, 56 and 57 depict this variation of current as a function of time for a short circuit at the terminals of a machine operating initially at no-load.

$$I_{\text{ac}} = E \left[\frac{1}{X_{\text{d}}} + \left(\frac{1}{X_{\text{d}}'} - \frac{1}{X_{\text{d}}} \right) e^{-t/T_{\text{d}}'} \right.$$

$$\left. + \left(\frac{1}{X_{\text{d}}''} - \frac{1}{X_{\text{d}}'} \right) e^{-t/T_{\text{d}}''} \right] \qquad \text{(Eq 55)}$$

$$I_{\text{dc}} = \frac{2 E \cos \alpha}{X''} e^{-t/T_{\text{d}}} \qquad \text{(Eq 56)}$$

$$I_{\text{T}} = \sqrt{I_{\text{ac}}^2 + I_{\text{dc}}^2} \qquad \text{(Eq 57)}$$

IEEE
Std 399-1980

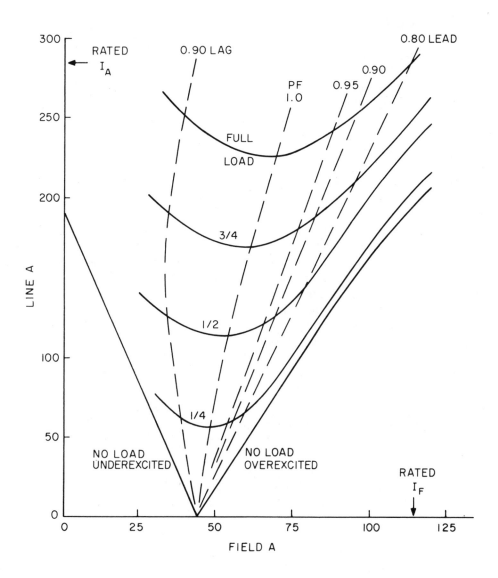

Fig 36
Vee–Curves: Synchronous Motor, 2000 hp,
4000 V, 180 r/min, 0.8 Lead Power Factor

The total current I is made up of two components:

(1) A power frequency ac component I_{ac}, the rms value of which decreases with time t, in accordance with Eq 55. This component is called the symmetrical current.

(2) A dc component I_{dc}, which decreases with time in accordance with Eq 56.

The initial magnitude of I_{dc} is a function of the angle α of the voltage wave at which the short circuit occurred. It is the amount of offset of the current wave. It is proportional to the rate of change of voltage at the instant of short-circuit. The offset is maximum at $\alpha = 0$, because the rate of change of voltage is maximum when the voltage is 0. The offset is 0 at $\alpha = 90°$ because the rate of change is zero at the positive or negative peak voltage values.

In the above equations:

E = the open circuit voltage

X_d, X_d', X_d'' = the direct-axis synchronous, transient, and subtransient reactances, respectively

T_a, T_d', T_d'' = the armature and the direct-axis transient and subtransient short-circuit time constants, respectively

For typical values of reactances and time constants, and with the maximum offset condition ($\cos \alpha = 1$) the equations will reduce to:

At time $t = 0$ (under subtransient conditions)

$$I_{ac}'' = \frac{E}{X_d''} \qquad \text{(Eq 58)}$$

$$I_{dc}'' = \frac{\sqrt{2}E}{X_d''} \qquad \text{(Eq 59)}$$

$$I_T'' = \frac{E}{X_d''}\sqrt{1+2} = \frac{E}{X_d''}\sqrt{3} \qquad \text{(Eq 60)}$$

At time $t = \infty$

$$I_{ac} = \frac{E}{X_d}$$
$$I_{dc} = 0 \qquad \text{(Eq 61)}$$
$$I_T = I_{ac} = \frac{E}{X_d}$$

Since T_d' is much larger than T_d'', there is a time t' smaller than T_d' and larger than T_d'', that will make $e^{-t/T_d'}$ approximately equal to 1 and $e^{-t/T_d''}$ approximately equal to 0. In this case, the original equations reduce to:

$$I_{ac}' = \frac{E}{X_d'} \qquad \text{(Eq 62)}$$

$$I_{dc}' = \frac{\sqrt{2}\,E\,e^{-t/T_a}}{X_d''} \qquad \text{(Eq 63)}$$

Should there be an impedance $Z_L = R_L + jX$ between the machine terminal (still at no load) and the point of fault, Eqs 58, 62 and 61 will become:

$$I_{ac}'' = \frac{E}{R_L + j(X_L + X_d'')} \qquad \text{(Eq 64)}$$

$$I_{ac}' = \frac{E}{R_L + j(X_L + X_d')} \qquad \text{(Eq 65)}$$

$$I_{ac} = \frac{E}{R_L + j(X_L + X_d)} \qquad \text{(Eq 66)}$$

The armature Time constant T_a will be shortened appreciably by addition of resistance R_L in the external circuit. So Eq 63 will become:

$$I_{dc}' = \frac{\sqrt{2}\,E\,e^{-t/T_a'}}{X''}_{(T_a' < T_a)} \qquad \text{(Eq 67)}$$

The offset current will decay more rapidly the farther away (electrically) the machine is from the point of fault.

The voltage E, in all the above equations, is equal to the terminal voltage V_t, since it was assumed that the machine was carrying no load before the short circuit. If the machine is carrying a current I_L before the short circuit, the voltage E will be different in each equation, to satisfy prefault conditions. For the case of a generator the voltage in Eqs 58, 62 and 61 will be:

$$E'' = V_t + I_L X_d'' \qquad \text{(Eq 68)}$$

$$E' = V_t + I_L X_d' \qquad \text{(Eq 69)}$$

$$E = V_t + I_L X_d \qquad \text{(Eq 70)}$$

respectively. These voltages have been called, voltage behind subtransient reactance (E''), voltage behind transient reactance (E'), and voltage behind synchronous reactance (E). It is not practical, in short circuit studies, to calculate the system currents for the entire period — from the time of fault to the time that the current reaches a steady state value. The normal procedure is to resolve the network at times $t = 0$ or $t = t'$, or both, using the models of Fig 37. Network solutions at $t = \alpha$ are meaningless since the machine field excitation has likely been changed at that time.

Depending on the study objectives, the effect of the offset current I_{dc} may or may not be important. In power circuit breaker applications, however, it is a very important consideration. To obviate the difficulties in resolving Eq 63, the breaker standards specify multipliers for the X_d'' and X_d' current components. These are a function of machine type and of the time from the inception of the short circuit. Figure 22 of [10] lists those multipliers, gives examples of their use and expands on this important aspect of short circuit studies.

The models of Figs 37 and 38 are also applicable to synchronous motors and synchronous condensers, the difference being that the E'', E' and E voltages are calculated with:

$$E'' = V_t - I_L X_d'' \qquad \text{(Eq 71)}$$

$$E' = V_t - I_L X_d' \qquad \text{(Eq 72)}$$

$$E = V_t - I_L X_d \qquad \text{(Eq 73)}$$

Fig 37
Models of Synchronous Machines for Short-Circuit Studies

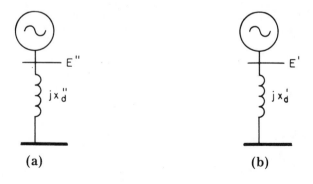

(a)   (b)

4.9.3.3 Stability Models. In stability studies it is necessary to resolve the electrical networks at a series of intervals starting from the time of inception of a disturbance and for as long as necessary to establish a trend in the system performance. This may cover a period shorter than 1 s for transient stability or several seconds for dynamic stability studies.

4.9.3.3.1 Classical Model. The classical model for transient stability study consists of a simple constant voltage source behind a constant transient reactance as in Fig 37(b). This model neglects the following factors by assuming that:

(1) The shaft mechanical power remains constant

(2) Field flux linkages remain constant

(3) Damping is non-existent

(4) The constant voltage and reactance are not affected by speed variations

(5) The rotor mechanical angle coincides with the phase angle of the internal voltage

A system found stable under these assumptions will likely be stable if any or all of the above factors are taken into consideration. However, should instability

Fig 38
General Model for AC Machines in Short-Circuit Studies

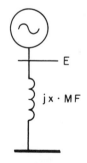

MF = multiplier specified
in ANSI/IEEE C37.010-1979

be indicated under the above assumptions, the system may in fact be stable! Thus there is a need to account for the neglected factors in certain circumstances. Which factors should be considered under which circumstances? This question will remain unanswered in this text. The various models, in increasing degree of sophistication, will be presented with simple explanations, if possible, or without explanations in which case references will be given for the reader to consult.

4.9.3.3.2 The H Constant. Stability studies are concerned with relative speed variations of rotating masses. The kinetic energy of a rotating mass, using units of Table 3, is:

$$KE = \tfrac{1}{2} I\omega^2 = \tfrac{1}{2} M\omega \text{ J} \qquad \text{(Eq 74)}$$

The rotating mass associated with a generator includes the rotor, shaft, coupling, turbine and exciter, if the rotating type is used. Since this mass is designed to rotate at a fixed (synchronous) speed, the stored energy, at synchronous speed, of a given machine is used as a constant. This constant is usually normalized by defining a quantity H that expresses the stored energy per unit of machine rated power. If kinetic energy is in megajoules and the rated power S in megavolt-amperes the H constant will be:

$$H = \frac{KE}{S} = \frac{J}{VA}$$

Since 1 joule is equal to 1 watt-second, the H constant is sometimes stated in equivalent kilowatt-second per kVA units.

To calculate the H constant use:

$$H = 0.231 \frac{(WR^2)\,(r/min)^2 \cdot 10^{-3}}{MVA} \qquad \text{(Eq 75)}$$

where WR^2 is the moment of inertia in pounds-feet squared and r/min is the speed in revolutions per minute. The H constant values will fall within the narrow range of approximately 1 to 15, irrespective of the size of the machines.

4.9.3.3.3 Stability Model Variations.
In the discussion of 4.9.3.2, only the direct-axis parameters were considered on the basis that short circuits produce currents of low power factor (quadrature currents predominate). This assumption may not be acceptable for disturbances considered in stability analysis. Therefore, additional synchronous machine parameters are required to more accurately model the behavior and account for the differences in the magnetic construction types, such as: salient poles, smooth rotor, laminated rotor, solid iron rotor with or without dampers.

Quadrature-axis reactances and open-circuit time constants are defined for that purpose. Chapter 1 of [11] is especially recommended as a clear and basic text in the subject.

The classical model may be improved one step by taking account of the variation of X_d' with time from its initial value to a steady state value of X_d. The variation will be an exponential described by a time constant T_{do}' (transient, open-circuit time constant). The three parameters X_d, X_d' and T_{do}' will ignore the major effect of dampers.

Another improvement will involve adding the effect of dampers which predominate during fast changing conditions, that is, the subtransient state. The additional parameters X_d'' and T_{do}'' will take care of this effect.

The saliency of the rotor will be represented by the quadrature-axis parameters, X_q, X_q' and X_q'' and the associated time constants T_{qo}', T_{qo}''.

Adding the parameters X_q' and T_{qo}' to those of the first improved model will increase the accuracy in the case of a generator with solid iron rotor. But, as before, the damping effect will have been mostly neglected. The solid iron rotor will be fully represented by all direct- and quadrature-axis, synchronous, transient and subtransient reactances and associated time constants.

The transient quadrature-axis reactance of salient-pole machines has the same value as the equivalent synchronous reactance. Thus salient-pole machines can be fully modeled as the solid-iron rotor machine by omission of the X_q' and T_{qo}' parameters.

4.9.3.4 Exciter Models.

(1) *Saturation.* The field poles saturate as the excitation current exceeds a certain level (see Fig 39). Computer programs will usually account for the related nonlinearity of air-gap voltage and field current from input data representing two points on the saturation curve. Refer to program instructions for which two points to use.

(2) *Standard Models.* An IEEE committee has developed a number of models to represent excitation systems and the dynamic characteristics of synchronous machines for stability studies [12]. A tutorial paper [13] supplements reference [12] by discussing the transfer function blocks and their practical meanings as well as other topics related to excitation system response. Only type (1) is repeated here (Fig 40) to illustrate the following points.

Consider the simple circuit of Fig 41. If the input voltage V is a step function (voltage changes suddenly at $t = 0$ from 0 to 1.0 pu V), the output voltage V will be an exponential function of time:

$$V_o = V_i\, e^{t/RC} \tag{Eq 76}$$

that may be rearranged

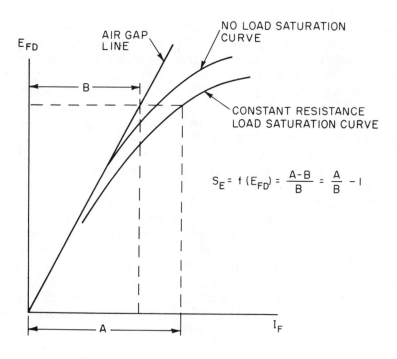

$$S_E = f(E_{FD}) = \frac{A-B}{B} = \frac{A}{B} - 1$$

Fig 39
Saturation Curves

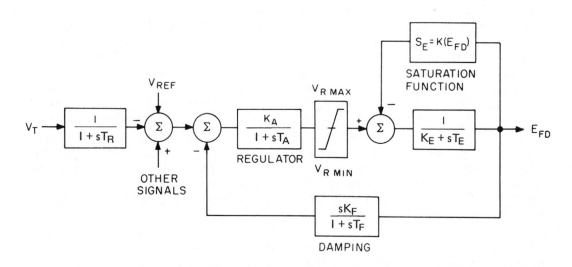

Fig 40
IEEE Type 1 Excitation System

86

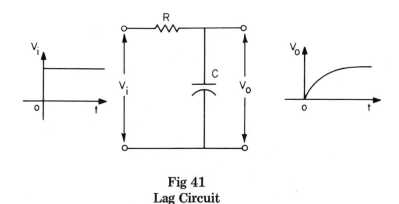

Fig 41
Lag Circuit

$$\frac{V_o}{V_i} = e^{t/RC} \qquad \text{(Eq 77)}$$

If the input is sinusoidal voltage, the output will also be sinusoidal but its amplitude will be reduced by the factor

$$\frac{1}{1 + j\omega CR}$$

and phase shifted by a lagging angle

$$\phi = \tan^{-1} \frac{R}{\omega C}$$

since:

$$V_o = V_i \frac{1}{1 + j\omega CR} \qquad \text{(Eq 78)}$$

where

$\omega = 2\pi f$ and f is frequency.

Dividing both sides of Eq 78 by V_i yields:

$$\frac{V_o}{V_i} = \frac{1}{1 + j\omega CR} \qquad \text{(Eq 79)}$$

This equation has the same form as the equation $1/(1 + sT_B)$ (transfer function) in the first block of the type 1 excitation system ($s = j\Omega$ and $T_B = RC$). Analysis of the response of this simple RC *lag* circuit can therefore be made in the time domain (Eq 76) or the frequency domain (Eq 78). It is common practice to use the frequency domain representation for both the excitation (as in Fig 40) and the prime mover systems. It *pictorializes* the system.

The lag transfer function just introduced has many applications, a few of which are representations of:

(1) The delay for the field current to change to a steady state value following a change of input field voltage

(2) The delay of water flow rate changes in penstocks

(3) The delays inherent in components of control mechanisms.

A similar analysis with the circuit of Fig 42 would yield the *lead* transfer function which has the form $s/(1 + sT)$. Its response time can be compared with the previous one. In the type 1 excitation system this function serves the purpose of damping the effect of the amplification K_A of the regulator that forces the field for faster response. K_F is the fraction of the output field voltage that is fed back to accomplish this damping.

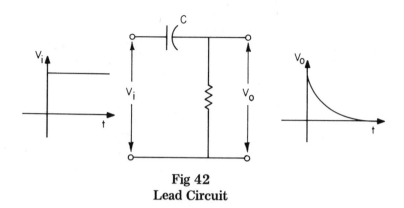

Fig 42
Lead Circuit

4.9.3.5 Prime Movers and Governor Models. Basic models for speed-governing systems and turbines in power system stability studies have been presented in an IEEE Committee Report [14]. As mentioned earlier, the models are in the form of block diagrams with transfer functions describing the system components' performance. Two more papers [15], [16] cover some of the basics and will help the novice to understand the relationships between the physical elements and the transfer functions.

Typical parameter values are also available in these references. Of course, the analyst will be well advised to seek from the manufacturer the data applicable to his equipment before compromising with typical data.

4.10 Miscellaneous Bus Element Models

4.10.1 Lighting and Electric Heating. Lighting and electric heating often constitute a large section of a plant load, particularly in commercial buildings. This type of load can be modeled as constant admittance as suggested by Figs 4.10, 4.11, and 4.12 of [17]. The constant admittance model would seem appropriate for fluorescent and mercury vapor lighting as well as for incandescent lighting. In utility type networks where substations are equipped with voltage regulators, lighting and heating can be represented as constant $P + jQ$.

To calculate the admittances determine the watts P and vars Q at rated voltage V and resolve

$$Y \text{ (siemens)} = \frac{P + jQ}{E^2}$$

The fluorescent and mercury vapor lighting power factors will be determined from manufacturers' data in order to calculate the vars Q. Incandescent lights and electric heating will have unity power factors.

4.10.2 Electric Furnaces. In load flow studies this load will usually be represented by constant power which will reflect a desired controlled operating condition to be analyzed. In the case of short circuit and stability studies the electric furnaces may behave like a constant impedance load. It is unlikely that automatic load tap changers and electrode position controls will have had time to change from the prefault condition to the end of the period covered in transient stability studies. This may very well be the case also in dynamic stability studies extending to several s. Since furnace loads can be a large section of a plant load, it is important that the constant impedance model be specified in stability problems because of its damping effect on the power system.

4.10.3 **Shunt Capacitors.** Banks of capacitors are used extensively to correct power factors and as a result, improve voltage regulation at the point of connection. They are modeled simply by a constant shunt capacitive susceptance.

$$jB = j2\pi fC$$

where C is in farads, f in hertz and B in siemens.

Often the capacitors will be specified in kvars. The rated voltage of the bus where the capacitors are connected must be known whether the susceptance or the equivalent kvars at the base voltage are used in the study. Remember that kvars vary proportionally as the square of the voltage for a constant susceptance.

4.10.4 **Shunt Reactors.** These are seldom used in industrial power systems. A constant admittance would be the correct model for reactors that do not saturate. The same voltage considerations as in the last paragraph would apply.

4.11 References

[1] *Electrical Transmission and Distribution Reference Book*, East Pittsburgh, PA, Westinghouse Electric Corporation, 1964.

[2] PUCHSTEIN, A.F., and LLOYD, T.C. *Alternating-Current Machines*, New York, John Wiley & Sons, 1974.

[3] McFARLAND, T.C., and VAN NOSTRAND, D. *Alternating-Current Machines*, New York, John Wiley & Sons, 1948.

[4] KENT, M.H., SCHMUS, W.R., Mc CRACKIN, F.A., and WHEELER, L.M. Dynamic Modeling of Loads in the Stability Studies, *IEEE Transactions Power Apparatus and Systems*, May 1969.

[5] ADLER, R.B., and MOSHER, C.C. Steady State Power Characteristics for Power System Loads, *IEEE Paper 70CP 706-PWR*, 1970.

[6] ILICITO, F., CEYHAN, A., and RUCKSTUHL, G. Behavior of Loads During Voltage Dips Encountered in Stability Studies — Field and Laboratory Tests, *IEEE Transactions Power Apparatus and Systems*, Nov/Dec 1972.

[7] IEEE COMMITTEE REPORT, System Load Dynamics — Simulation Effects and Determination of Load Constants, *IEEE Transactions Power Apparatus and Systems*, Mar/Apr 1973.

[8] FITZGERALD, A.E., and KINGSLEY, Jr, C. *Electric Machinery*, New York, McGraw-Hill, 1961.

[9] ANSI/IEEE Std 112-1978, Standard Test Procedure for Polyphase Induction Motors and Generators.

[10] HUENING, Jr, W.C. Interpretation of New American National Standards for Power Circuit Breaker Applications, *IEEE Transactions Industry and General Applications*, Sept/Oct 1969.

[11] KIMBARK, E.W. Power System Stability: Synchronous Machines, vol 3, New York, Dover Publications, Inc, 1968.

[12] IEEE COMMITTEE REPORT, Computer Representation of Excitation Systems, *IEEE Transactions Power Apparatus and Systems*, June 1968.

[13] IEEE COMMITTEE REPORT, Excitation System Dynamic Characteristics, *IEEE Transactions Power Apparatus and Systems*, Jan/Feb 1973.

[14] IEEE COMMITTEE REPORT, Dynamic Models for Steam and Hydro Turbines in Power System Studies, *IEEE Transactions Power Apparatus and Systems*, Nov/Dec 1973.

[15] EGGENBERGER, M.A. A Simplified Analysis of the No–Load Stability of Mechanical–Hydraulic Speed Control Systems for Steam Turbines, *ASME Paper 60-WA-34.*

[16] RAMEY, D.G., and SKOOGLUND, J.W. Detailed Hydrogovernor Representation for System Stability Studies, *IEEE Transactions Power Apparatus and Systems*, Jan 1970.

[17] *Industrial Power Systems Handbook*, BEEMAN, D., Editor, New York, McGraw-Hill, 1955.

[18] *Aluminum Electrical Conductor Handbook*, The Aluminum Association, 750 Third Ave, New York, 1971.

[19] STEVENSON, Jr, W.D. *Elements of Power System Analysis*, New York, McGraw-Hill, 1975.

[20] *Standard Handbook for Electrical Engineers*, FINK and CARROL, Editors, New York, McGraw-Hill, 11th Edition.

[21] Industrial Power Systems Data Book, Schenectady, N.Y., General Electric Company.

[22] ANSI/IEEE C37.010-1979, IEEE Application Guide for AC High Voltage Circuit Breakers Rated on a Symmetrical Current Basis (Consolidated edition).

[23] ANSI/IEEE C37.5-1979, IEEE Guide for Calculation of Fault Currents for Application of AC High-Voltage Circuit Breakers Rated on a Total Current Basis (Revision of ANSI C37.5-1969).

5. Load Flow Studies

5.1 Introduction. Load flow is the terminology applied to the flow of power from one or more sources through available paths to loads consuming energy. Direction and amount of power flowing in each path or branch can be shown on a system map commonly referred to as the system single line (or one-line) diagram — a simplified visual model of a *balanced* three-phase electrical system.

When the system is radial and has no parallel paths, power flows directly to the load. Most systems today, however, are much more complex and have many paths or branches over which power can flow. Such a system forms a network of series and parallel paths.

Electric power flow in a network, like water flow in a complex water system, divides the flow among branches according to their respective impedances until a pressure or voltage balance is reached in accord with Kirchhoff's laws.

As long as the circuit remains unchanged, the balanced conditions hold. The flow will shift however, any time the circuit configuration is changed or modified, generation is shifted, or load requirements change. In any practical operating system, load distribution shifts each time a user turns a light, a motor, or other power consuming device on or off. Consequently, load flow is a fluid and ever-changing thing. Where parallel paths or circuits exist to supply power loads, the operation of switches, breakers, etc (whether manually by an operator or automatically by relay action) will change the circuit configuration and cause a redistribution of power flow through interconnecting lines. Large industrial plants and electric utility companies with complex system networks face a difficult task. They must provide operators and dispatchers with information necessary to ensure efficient operation, minimize losses, maintain reliability of service, and coordinate protective relaying for unexpected and emergency conditions.

In addition, the power system planner must look at future power requirements and allow for additions and changes. It is necessary to meet projected loads when they occur, and still maintain or improve the efficiency and reliability of the service.

For many years the only way to obtain data to guide operators and dispatchers was from experience gained from emergency outages or experimentation, or both. This method was often costly and sometimes drastic, and was unsatisfactory for future planning in that effects of future heavy loads could not be accurately determined without slow and tedious calculations.

Computer programs to solve load flows are divided into two types — static and dynamic (real time). Most load flow studies for system analysis are based on static network conditions. Real time load flows (on line) are used primarily as operating tools for optimization of generation, var control, dispatch, losses, and tie line control. This discussion is concerned with only the static network and its analysis. Two types of programs are available to users, time share versions and full-fledged batch programs. Where minimal input and output give a satisfactory analysis, time share programs are of value, particularly when study time is important. Usually thorough analysis requires more detailed input and a complete output report. Generally, entering data and printing a large output report is not feasible on a time share basis unless the terminal is equipped with card reading and high-speed printing equipment.

Since the advent of analog network analyzers and digital computers, with their ability to quickly solve load flow problems, the picture has changed radically. Now, starting with a system operating under normal conditions, the flows in all branches can be quickly computed for comparison with all other cases, present and future. It is simple to introduce one or more changes to the present normal system and calculate the effect on all its buses and branches.

Some changes that can be introduced individually, or in combination, to determine effect on the system are listed below:

(1) Take any line or transformer out of service

(2) Add load to any or all buses

(3) Change regulated bus voltages

(4) Add new transmission lines

(5) Remove, add or shift generation to any bus

(6) Change transformer taps

(7) Increase conductor size on lines

(8) Add rotating or static var supply to buses

(9) Control reactive power distribution

(10) Increase or decrease size of transformers

(11) Change MW or Mvar constraints, or both, between subnetworks

System changes can be simulated and results of such changes analyzed without jeopardizing operations or equipment. Each case study is accurate and reliable for conditions given. Analysis of the results of a case study often shows conditions which need further study. This can show elements of the system that should be changed to improve performance.

5.2 System Representation. Present utility and industrial plant electrical systems can be extensive. A simplified visual means of representing the complete system is essential to understand the operation of the system under its various possible operating modes. The system single line (or one-line) diagram serves this purpose. The single line diagram consists of a drawing, or sketch, identifying buses and interconnecting lines (see Fig 46). Loads, generators, transformers, reactors, capacitors, etc, are all shown in their respective places in the system. It is necessary to show equipment parameters as well as their relationship to each other.

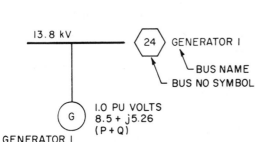

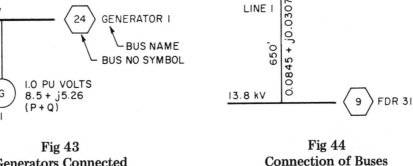

Fig 43
Generators Connected
to their Bus

Fig 44
Connection of Buses

Buses can be named as well as numbered. Interconnecting lines are shown with their $R + jX$ values entered or cross referenced with tables of values. For instance, generators are shown connected to their bus as illustrated in Fig 43 with their equipment parameters specified. Each line originates on a bus and terminates on a different bus as depicted in Fig 44. Line 1 runs from bus 3 (main 1)

to bus 9 (FDR 31) and is shown to be 650 ft long with $R = 0.0845\ \Omega$ and $X = 0.03074\ \Omega$.

Transformers are usually shown between two buses with the primary on one bus and the secondary winding identified by an auxiliary bus (identified by number and a suitable name, see Fig 45). The auxiliary bus can then be tied by a line to the other bus.

Fig 45
Auxiliary Bus

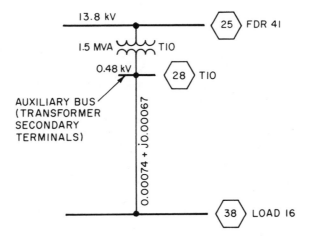

If the auxiliary bus is used, transformer impedance is entered along with line impedance. Otherwise, their series combination is entered. If the transmission or distribution line is long (5 mi or more), it will require charging (charging conductance G, and line leakage susceptance B). One half of the total for the line is usually shown on each of the two buses.

If MVA capability of the line is known, it can be entered also. Tap voltages, where required, should be entered to modify the nominal ratios. If transformers are of the tap changing under load type, the tap limits and incremental tap values should be given. If transformer impedance is given in percent (or per unit), it is expressed on its own self-cooled MVA rating as base.

5.3 System Data Organization. A load flow *solution* gives the power flow in all branches for a given set of conditions. It represents a steady state in which the influential parameters are in balance and a solution has been found. A load flow *study* is a series of such calculations made when certain equipment parameters are set at different values, or circuit configuration is changed by operning or closing breakers, adding or removing a line, etc.

Many load flow programs have been written for digital machines. They differ in some ways, mainly by solution techniques and sophistication (taking into account more and more influencing parameters). Power flow is very sensitive to circuit changes of impedances, interconnections, and location of loads and equipment. In general, less sophisticated programs will give satisfactory results, although more sophisticated ones will have increased accuracy. As programs become more sophisticated additional input data is required. Utility companies with large diverse systems tend to use the more sophisticated programs. Industrial plants that use less sophisticated programs find them satisfactory.

The balance of this section uses one of the less sophisticated programs to illustrate a load flow study for an industrial plant. Programs are available with various levels of sophistication. This one was arbitrarily chosen for convenience. The following is a case study for a large industrial plant and illustrates in detail how a load flow study is undertaken.

5.4 Load Flow Study Example

5.4.1 General. To illustrate the use of a load flow program a typical industrial plant will be studied. The single line connection diagram of the plant electrical system is shown in Fig 46. The base case solves the system in the normal operating mode supplying present maximum loads. A case identified as case A1 is solved to show what happens to the system when lightning causes relays to open breakers at bus 3 and bus 26 to isolate transmission line 3.

The impedance diagram in Fig 47 is similar to the single line diagram except that impedances of the interconnecting lines, equipment parameters, load requirements, and nominal bus voltages have been added.

5.4.2 Input Requirements. The following input data apply specifically to this program, but similar data will be required for almost any load flow program. Three data input forms are used for entering required data in the proper format for this particular load flow program.

The input data sheet shown in Fig 48 is used to identify the company, location, date, and title of the study, and who made the study. Also included is an unlimited number of statements defining power system conditions under which

IEEE
Std 399-1980

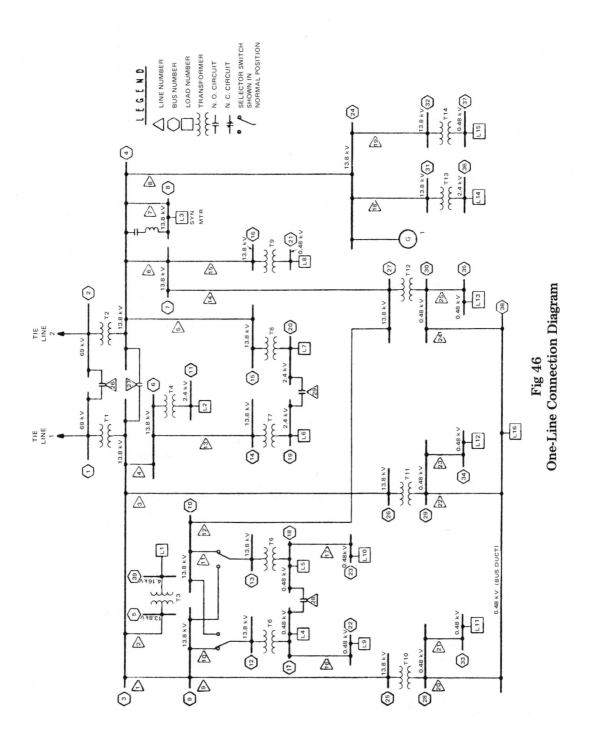

Fig 46
One-Line Connection Diagram

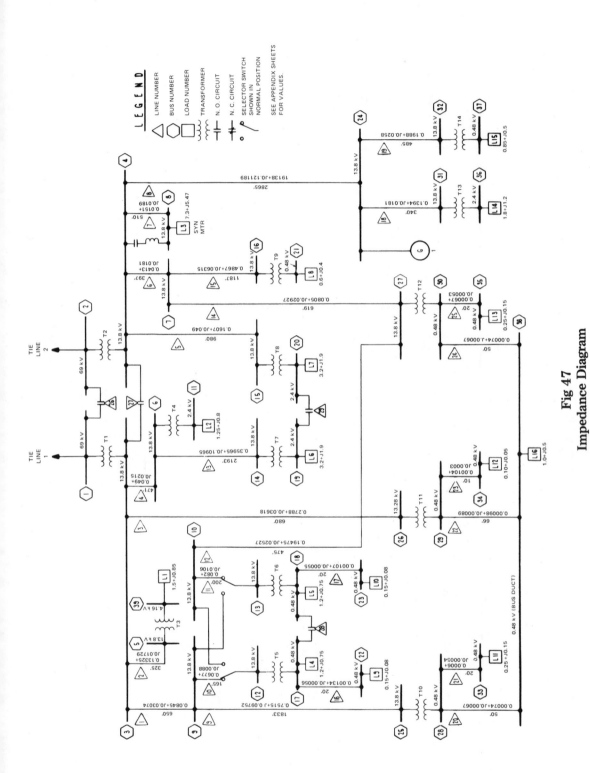

Fig 47

Impedance Diagram

DP.E113.1

JOB NUMBER 47-1333
CUSTOMER I&CPS
DATE 4/3/75

A B C CORPORATION
PROGRAM E113 - POWER FLOW

USER J.R.SMITH
LOCATION S211   EXT.2511
SHEET 1 OF 5

IDENTIFICATION AND CASE DESCRIPTION

- CASE NO. BASE - TYPICAL INDUSTRIAL PLANT ELECTRICAL SYSTEM
- ABC CORPORATION POWER FLOW PROGRAM E113
- JOB NO. 47/333 CUSTOMER I&CPS ANALYSIS SUBCOMMITTEE DATE 4/3/75
- PROJECT STANDARDS PUBL. #399 LOCATION ANYPLACE, USA BY J.R.SMITH
- TYPICAL EXAMPLE OF LOAD FLOW STUDY - BASE CASE - MAXIMUM 1974 LOADS
- SYSTEM CONFIGURATION - NORMAL

1st line will be used as title. Any no. of ** cards may be used for comment.

PROGRAM CONTROL CARDS
PUNCH ONLY AS REQUIRED

- COMPUTE SYSTEM MAXIMUM ITERATIONS AT PLACE - SYSTEM DATA AT END OF INPUT DATA AT EACH CASE DATA A
- C END NEW MAXIMUM ITERATIONS REAL, IMAGINARY DATA
- ACCELERATING FACTOR ENTER - 1ST - 2ND DESIRED IF 400 NOT DESIRED
- BUS SELECT - SELECTED BUSES BELOW LAST BUS MUST = 0

NOTE: Indicate number of control cards in required column.
These cards must be placed as required by program. (See Users Manual)

NO. REQUIRED
2
1

Fig 48
Input Data Sheet Form 1

calculations were made. Reference to Fig 48 gives an example of what is required.

The lower section of Fig 48 is used as a matter of convenience to obtain the required number of control cards needed to adapt the system to the program. This requirement is discussed later in the text.

Computer control to properly read and process input data is effected in this program by a *tag* or code number entered on input data in Figs 49 and 50. The tag identifies the data and ensures its proper interpretation and storage when read into the computer. *Tag* designations are as follows:

Tag	Description
0 (or blank)	Bus data card
1	Line data card
2	Transformer data card
3	Phase shift transformer
4	Tap changer under load
5	Low or negative reactance line

Figure 49 is used for entering all data relating to system buses, that is, bus name, bus number, bus type (see listing at bottom), nominal bus voltage and angle, generation, loads, and associated capacitors or reactors.

Figure 49 shows the information filled in with data for the load flow study example.

Bus types are used to direct the computer to properly process bus data. Types of buses are as follows:

Bus Type	Function
0	*Slack or swing bus.* Takes the swings in loads and adjusts to supply MW and Mvar losses.

1	*Unregulated active.* Buses that supply loads and which normally have no voltage regulation equipment.
2	*Regulated.* Buses on which specified voltage is held by supplying necessary vars (by some kind of suitable equipment such as generators, capacitors, reactors, etc).
3	*Var Limited.* A bus that can supply vars to hold the voltage only to specified var limits. (Similar to type 2 except with limits.)
4	*Floating.* A bus that due to switching, or error, is disconnected from the original system. (An automatic classification.)
5	*Passive.* A bus that retransmits all power received (no direct loads on bus).

Figure 50 is used for entering line and transformer data for the system, that is, *from* and *to* bus numbers, conductance G and susceptance B (usually neglected for in-plant distribution lines), resistance and reactance of the line or transformer, MVA rating, and transformer tap voltages.

Figure 50 shows data for the illustrative example.

Most programs allow changes to be made to a base case with a minimum of input data. Change data is entered on Fig 49 for bus changes, and in Fig 50 for line and transformer changes. A change code number must be entered in the proper column on the figures to instruct the computer of the change. The program will recognize the following changes to a base case:

DP.E113.2

JOB NUMBER 47/333
CUSTOMER I&CPS
DATE 4/3/75

A B C CORPORATION
PROGRAM E113 – POWER FLOW – BUS DATA

USER J.R.SMITH
LOCATION S2LL **EXT.** 2521
SHEET 2 OF 5

BUS NAME	CHANGE	TAG	BUS NO.	TYPE	KV BASE	BUS ANGLE	BUS P.U. VOLTS	GEN MEGA WATTS	GEN MEGA VARS	LOAD MEGA WATTS	LOAD MEGA VARS	MAX GEN MVARS	MIN GEN MVARS	CAPS REAC MVAR	MISC CODE	CASE NO.
TIE 1			1	3	690		10000	1000	600			1000	100			BASE
TIE 2			2	0.	690		10000	2500	1500			1500	100			
MAIN 1			3	5	138		10000									
MAIN 2			4	5	138		10000									
FDR32			5	5	138	10	9950									
FDR34			6	5	138	10	9950									
FDR42			7	5	138	10	9950									
SYNMO			8	1	138	10	9950			720	-540					
FDR31			9	5	138	100	9950									
FDR271			10	5	138	100	9800									
LOAD2			11	1	24	100	9400			125	80					
FDR92			12	5	138	15	9900									
FDR101			13	5	138	250	9800									
FDR61			14	5	138	20	9900									
FDR41			15	5	138	10	9900									
FDR72			16	5	138	15	9900									
TX5			17	1	.48	250	9750			120	75					
TX6			18	1	.48	250	9750			120	75					
LOADG			19	1	24	200	9800			320	190					
LOAD7			20	1	24	200	9850			320	190					

CODES:

CHANGE
0 NEW BUS
1 NEW BUS VOLTS
2 NEW GENERATION
3 NEW LOAD
4 NEW VAR LIMITS
5 NEW CAPS/REAC.

TYPE
0 SLACK
1 UNREGULATED ACTIVE
2 REGULATED
3 VAR LIMITED
4 FLOATING
5 PASSIVE
6 PHASE SHIFTER

TAG
0 BUS CARD
1 LINE CARD
2 TRANSFORMER CARD
3 PHASE SHIFT TRANSFORMER
4 TAP CHANGER UNDER LOAD
9 TRAILER CARD

Fig 49
Input Data Sheet Form 2

DP.E113.3

A B C CORPORATION

PROGRAM E113 – POWER FLOW – LINE/TRANSFORMER DATA

JOB NUMBER 471333
CUSTOMER I&CPS
DATE 4/3/75

USER J.R.SMITH
LOCATION 521L   EXT. 252L
SHEET 4 OF 5

FROM BUS ii	TO BUS jj	CIRCUIT	CHANGE	TAG	COND. G_ii%	SUSCEPT. B_ii%	RESISTANCE R_jj%	REACTANCE X_jj%	COND. G_jj%	SUSCEPT. B_jj%	LINE MVA RATING	AREA	TAP_ii	TAP_jj	BASE_ii	BASE_jj	BASE MVA	CASE NO.
3	9			1			4437	1614										BASE
3	5			1			6997	908										
3	26			1			14640	1900										
3	6			1			2572	1130										
4	15			1			8439	2573										
4	7			1			2168	953										
4	8			1			759	991										
4	24			1			10050	6364										
9	25			1			39463	5121										
9	12			1			3552	461										
10	13			1			4306	559										
10	27			1			10226	1327										
6	14			1			18885	5758										
7	16			1			4226	1537										
7	27			1			25555	3316										
17	22			1			57986	24392										
18	23			1			46354	23958										
24	31			1			7392	950										
24	32			1			10442	1355										
28	38			1			32227	29188										

1. Shunt conductance to ground – (corona, leakage, etc.) put 1/2 on each bus.
2. Shunt susceptance (capacitance to ground) enter 1/2 on each bus.

CODES:

CHANGE
0 NEW LINE OR TRANSFORMER
1 NEW TAPS—COLUMNS 49–64
2 REMOVE L/T
3 RESTORE L/T
4 NEW MVA RATING

TAG
0 BUS CARD
1 LINE CARD
2 TRANSFORMER CARD
3 LOW OR NEGATIVE REACTANCE
9 TRAILER CARD

AREA
0 NO LOSS SUMMATION
1 SUM LOSSES

Fig 50
Input Data Sheet Form 3

Bus Changes

Change Code	Description
0	Add new bus
1	Change bus volts
2	Add new generation
3	Add new loads
4	Change var limits
5	Add new capacitors or reactors

Line Changes

Change Code	Description
0	New line
1	New transformer taps
2	Remove line/transformer
3	Restore line/transformer
4	New line/transformer MVA rating

5.4.3 Special Data. Data can be entered to change the normal built-in constants if desired. This program uses the Gauss-Sidel iteration method of solution and involves starting the calculations with nominal or input values of voltages as specified on input data sheets, calculating the flow of currents through the branches, checking voltage drops in the branches and comparing the resultant voltage at each bus with the bus voltage. If the voltages are not within an acceptable tolerance, bus voltages are modified by an *acceleration factor* (real and imaginary) and calculations are repeated until they are within an acceptable tolerance. Each recalculation is called an iteration. If an entry is not made in the special data input sheet for *iterations*, the program will do 500 before aborting. This number can be overriden by entering the desired number on the form. Built-in acceleration factors are 1.6 for real values and 1.7 for imaginary values. These can be overriden by entering the preferred factors on the input form.

Normally the built-in values find a satisfactory solution. Optimum values give a solution with the fewest iterations. Different networks have different optimum values. Trial and error selection is the only practical way to determine the best values. These once found for the network, will continue to be near optimum unless the network is changed quite substantially. If only specified branch or bus flows are printed rather than a complete system output, the desired buses can be entered on Fig 48 under *Bus Select*. Up to 15 buses can be specified. Other special cards are used to control the computer run. A *Compute* card must be placed at the end of a case run to instruct the computer that all case data have been read and to start load flow calculations. Case studies can be stacked when changes to the first case are desired. Place the change cards after the *Compute* card and terminate the change case with another *Compute* card. Any number of changes can be stacked. Two different systems can be stacked by placing a *New System* card after the last *Compute* card, and giving full system data in the cards that follow and again terminating with a *Compute* card. After data for the last case are terminated, an *End* card should follow to instruct the computer that all desired work has been done.

5.5 Input Card Preparation. Having obtained and entered the system data in the input forms, it is necessary to have the data punched into computer input cards. For in-house computer systems, a card keypunch and operator are available to punch the data into cards and verify them. For in-house work, cards are arranged in the order specified in the user's manual (supplied with the program being used) and submitted to the computer center for processing. After processing, the results will be returned (with the input cards) to the user. For remote terminal entry, the computer service usually has keypunch machines and operators, who can punch cards and verify input data. The computer service will also assist in entering the cards by remote terminal reader and will notify the user when output results are ready. In some instances, data can be punched into paper or magnetic tape and submitted. However, punched cards are most frequently used. For the less experienced, there are consultants who can do the analysis and present the user with a complete report including a technical analysis of the computer output and suggestions and advice on system improvement.

5.6 Load Flow Results. Output from the computer for each case is printed in report form. For input data checking, a page of *input data* is printed from the computer file for all lines and transformers. The number of iterations the computer required is also printed (see Fig 51).

If a satisfactory solution cannot be obtained in the specified number of iterations (500), a message is printed stating that the number of specified iterations has been exceeded. In this event, the full report is printed, but caution should be used, as the values printed do not give a satisfactory balanced solution. The full report is often quite helpful in analyzing the system trouble that might have led to the unsatisfactory solution.

Assuming a solution has been reached, the report will be printed on as many sheets as required. Printed on the first line are the case number and description for identification purposes. Reference to Fig 52 verifies the following: The lowest

numbered bus is shown with name, bus number and voltage at solution, and phase angle with respect to the swing bus. If the bus is a load bus, load MW and Mvar are printed. Printed next is the bus number of the *From* bus, then the *To* bus number and name, MW, Mvar, and MVA flowing in the line. A plus (+) indicates flow of power from the first bus to the second bus. A negative sign (−) indicates the flow to be in the opposite direction. Note that MW flow and Mvar flow can be in opposite directions in the same line. This is not a desirable condition, but is entirely possible in real systems. For each line the MW and Mvar line loss is printed. All lines connected to this bus are given. If a transformer connects the two buses, the transformer tap voltages used in the calculations are printed. Finally, the bus error, or mismatch, is printed. When all buses with inter-connecting lines have been printed, the total accumulated bus error is printed. The last item printed is the total loss for the system. There are instances where a solution is found for specified conditions without exceeding the iteration limit, but does have excessive bus error or mismatch (power into the bus does not match power out of the bus, within a reasonable tolerance). When this occurs, voltage tolerance should be decreased for a more accurate solution. The amount of mismatch is a good indicator of the validity of the solution.

Some programs print certain values (usually bus voltages) in per unit values rather than actual. This method allows a relative evaluation. Multiplication by the per unit value permits determination of actual values. If the program does not reach a satisfactory solution (as evidenced by exceeding the specified number of iterations) the reason can be because of incomplete or inconsistent

INPUT DATA

** CASE NO. BASE - TYPICAL INDUSTRIAL PLANT ELECTRICAL SYSTEM

** CASE NO. BASE - TYPICAL INDUSTRIAL PLANT ELECTRICAL SYSTEM

**ABC CORPORATION   POWER FLOW   PROGRAM E113

** JOB NO. 471333   CUSTOMER IECPS ANALYSIS SUBCOMMITTEE   DATE 4/3/75

** PROJECT STANDARDS PUBL. #399   LOCATION ANYPLACE, USA   BY J.R.SMITH

** TYPICAL EXAMPLE OF LOAD FLOW STUDY - BASE CASE - MAXIMUM 1974 LOADS

** SYSTEM CONFIGURATION - NORMAL

III	JJJ	C	CHG	TAG	G(III)	B(III)	R	X	G(JJ)	B(JJ)	LNMVA	A	TAPI	TAPJ	BASEI	BASEJ	MVA
3	9	0	0	1	0.0	0.0	4.437	1.614	0.0	0.0	0.0	0	0.0	0.0	0.0	0.0	0.0
3	5	0	0	1	0.0	0.0	6.997	0.908	0.0	0.0	0.0	0	0.0	0.0	0.0	0.0	0.0
3	26	0	0	1	0.0	0.0	14.640	1.900	0.0	0.0	0.0	0	0.0	0.0	0.0	0.0	0.0
3	6	0	0	1	0.0	0.0	2.572	1.130	0.0	0.0	0.0	0	0.0	0.0	0.0	0.0	0.0
4	15	0	0	1	0.0	0.0	8.439	2.573	0.0	0.0	0.0	0	0.0	0.0	0.0	0.0	0.0
4	7	0	0	1	0.0	0.0	2.168	0.953	0.0	0.0	0.0	0	0.0	0.0	0.0	0.0	0.0
4	8	0	0	1	0.0	0.0	0.759	0.991	0.0	0.0	0.0	0	0.0	0.0	0.0	0.0	0.0
4	24	0	0	1	0.0	0.0	10.050	6.364	0.0	0.0	0.0	0	0.0	0.0	0.0	0.0	0.0
4	25	0	0	1	0.0	0.0	39.463	5.121	0.0	0.0	0.0	0	0.0	0.0	0.0	0.0	0.0
9	12	0	0	1	0.0	0.0	3.552	0.461	0.0	0.0	0.0	0	0.0	0.0	0.0	0.0	0.0
10	13	0	0	1	0.0	0.0	4.306	0.559	0.0	0.0	0.0	0	0.0	0.0	0.0	0.0	0.0
10	27	0	0	1	0.0	0.0	10.226	1.327	0.0	0.0	0.0	0	0.0	0.0	0.0	0.0	0.0
4	14	0	0	1	0.0	0.0	18.865	5.758	0.0	0.0	0.0	0	0.0	0.0	0.0	0.0	0.0
7	16	0	0	1	0.0	0.0	4.226	1.537	0.0	0.0	0.0	0	0.0	0.0	0.0	0.0	0.0
17	22	0	0	1	0.0	0.0	25.552	3.316	0.0	0.0	0.0	0	0.0	0.0	0.0	0.0	0.0
18	23	0	0	1	0.0	0.0	57.986	24.392	0.0	0.0	0.0	0	0.0	0.0	0.0	0.0	0.0
24	31	0	0	1	0.0	0.0	46.354	23.958	0.0	0.0	0.0	0	0.0	0.0	0.0	0.0	0.0
24	32	0	0	1	0.0	0.0	7.392	0.950	0.0	0.0	0.0	0	0.0	0.0	0.0	0.0	0.0
28	38	0	0	1	0.0	0.0	10.442	1.355	0.0	0.0	0.0	0	0.0	0.0	0.0	0.0	0.0
26	36	0	0	1	0.0	0.0	32.227	29.188	0.0	0.0	0.0	0	0.0	0.0	0.0	0.0	0.0
30	34	0	0	1	0.0	0.0	25.781	23.351	0.0	0.0	0.0	0	0.0	0.0	0.0	0.0	0.0
30	38	0	0	1	0.0	0.0	42.539	38.529	0.0	0.0	0.0	0	0.0	0.0	0.0	0.0	0.0
30	35	0	0	1	0.0	0.0	45.139	12.891	0.0	0.0	0.0	0	0.0	0.0	0.0	0.0	0.0
2	3	0	0	1	0.0	0.0	32.227	29.188	0.0	0.0	0.0	0	0.0	0.0	0.0	0.0	0.0
2	4	0	0	1	0.0	0.0	33.420	23.177	0.0	0.0	0.0	0	0.0	0.0	0.0	0.0	0.0
5	39	0	0	2	0.0	0.0	1.000	8.000	0.0	0.0	15.0	0	69.0	13.8	69.0	13.8	15.0
11	11	0	0	2	0.0	0.0	1.000	8.000	0.0	0.0	15.0	0	69.0	13.8	69.0	13.8	15.0
12	17	0	0	2	0.0	0.0	0.800	6.000	0.0	0.0	1.7	0	13.8	4.2	13.8	4.2	1.7
14	18	0	0	2	0.0	0.0	0.700	5.500	0.0	0.0	1.5	0	13.8	2.4	13.8	2.4	1.5
15	19	0	0	2	0.0	0.0	0.800	6.750	0.0	0.0	1.5	0	13.8	0.5	13.8	0.5	1.5
15	20	0	0	2	0.0	0.0	0.800	6.750	0.0	0.0	3.7	0	13.8	2.4	13.8	2.4	3.7
16	21	0	0	2	0.0	0.0	0.900	5.500	0.0	0.0	3.7	0	13.8	2.4	13.8	2.4	3.7
25	23	0	0	2	0.0	0.0	0.700	5.500	0.0	0.0	0.8	0	13.8	0.5	13.8	0.5	0.8
26	29	0	0	2	0.0	0.0	0.800	5.750	0.0	0.0	1.5	0	13.8	0.5	13.8	0.5	1.5
27	30	0	0	2	0.0	0.0	0.300	5.750	0.0	0.0	1.5	0	13.8	0.5	13.8	0.5	1.5
31	36	0	0	2	0.0	0.0	0.800	5.750	0.0	0.0	1.5	0	13.8	0.5	13.8	0.5	1.5
32	37	0	0	2	0.0	0.0	0.800	5.750	0.0	0.0	2.5	0	13.8	2.4	13.8	2.4	2.5
33	27	0	0	2	0.0	0.0	0.800	5.750	0.0	0.0	1.0	0	13.8	0.5	13.8	0.5	1.0

TOTAL ITERATION = 329

Fig 51

Printed Computer Output

```
**  CASE NO. BASE - TYPICAL INDUSTRIAL PLANT ELECTRICAL SYSTEM

TIE 1     1   69.00 KV      7.3 DEGREE
                            GENERATION    10.00 MW      5.79 MVAR   SCHEDULED BUS VOLTAGE    69.00
      1-   3 0   MAIN 1     10.00 MW       5.79 MVAR     11.56 MVA  LOSSES =    0.09 MW    0.71 MVAR
                            TRANSFORMER TAP  69.00 KV    13.80 KV
                            BUS ERROR       -0.00 MW    -0.00 MVAR


TIE 2     2   69.00 KV      0.0 DEGREE
                            GENERATION     5.32 MW       2.16 MVAR
      2-   4 0   MAIN 2      5.32 MW        2.16 MVAR      5.75 MVA  LOSSES =    0.02 MW    0.18 MVAR
                            TRANSFORMER TAP  69.00 KV    13.80 KV
                            BUS ERROR        0.00 MW     0.00 MVAR


MAIN 1    3   13.30 KV      4.4 DEGREE
      3-   1 0   TIE 1      -9.91 MW       -5.03 MVAR     11.14 MVA  LOSSES =    0.09 MW    0.71 MVAR
                            TRANSFORMER TAP  13.80 KV    69.00 KV
      3-   5 0   FDR32       1.52 MW        0.98 MVAR      1.81 MVA  LOSSES =    0.00 MW    0.00 MVAR
      3-   6 0   FDR34       4.54 MW        3.07 MVAR      5.48 MVA  LOSSES =    0.01 MW    0.00 MVAR
      3-   9 0   FDR31       2.61 MW        0.96 MVAR      2.78 MVA  LOSSES =    0.00 MW    0.00 MVAR
      3-  26 0   FDR33       1.24 MW        0.04 MVAR      1.24 MVA  LOSSES =    0.00 MW    0.00 MVAR
                            BUS ERROR        0.00 MW     0.03 MVAR


MAIN 2    4   13.60 KV     -1.6 DEGREE
      4-   2 0   TIE 2      -5.30 MW       -1.99 MVAR      5.66 MVA  LOSSES =    0.02 MW    0.18 MVAR
                            TRANSFORMER TAP  13.80 KV    69.00 KV
      4-   7 0   FDR42       1.14 MW        2.47 MVAR      2.72 MVA  LOSSES =    0.00 MW    0.00 MVAR
      4-   8 0   SYNM0       7.15 MW       -5.40 MVAR      8.96 MVA  LOSSES =    0.01 MW    0.01 MVAR
      4-  15 0   FDR41       3.24 MW        2.14 MVAR      3.89 MVA  LOSSES =    0.01 MW    0.00 MVAR
      4-  24 0   GEN 1      -6.14 MW        2.73 MVAR      6.74 MVA  LOSSES =    0.05 MW    0.03 MVAR
                            BUS ERROR       -0.10 MW    -0.01 MVAR


FDR32     5   13.28 KV      4.4 DEGREE
      5-   3 0   MAIN 1     -1.52 MW       -0.98 MVAR      1.81 MVA  LOSSES =    0.00 MW    0.00 MVAR
      5-  39 0   LOAD1       1.52 MW        0.97 MVAR      1.80 MVA  LOSSES =    0.02 MW    0.12 MVAR
                            TRANSFORMER TAP  13.80 KV     4.16 KV
                            BUS ERROR        0.00 MW     0.01 MVAR


FDR34     6   13.26 KV      4.4 DEGREE
      6-   3 0   MAIN 1     -4.53 MW       -3.06 MVAR      5.47 MVA  LOSSES =    0.01 MW    0.00 MVAR
      6-  11 0   LOAD2       1.26 MW        0.89 MVAR      1.55 MVA  LOSSES =    0.01 MW    0.09 MVAR
                            TRANSFORMER TAP  13.80 KV     2.40 KV
      6-  14 0   FDR61       3.27 MW        2.16 MVAR      3.92 MVA  LOSSES =    0.03 MW    0.01 MVAR
                            BUS ERROR       -0.00 MW     0.01 MVAR


FDR42     7   13.59 KV     -1.5 DEGREE
      7-   4 0   MAIN 2     -1.14 MW       -2.47 MVAR      2.72 MVA  LOSSES =    0.00 MW    0.00 MVAR
      7-  16 0   FDR72       0.61 MW        0.45 MVAR      0.75 MVA  LOSSES =    0.00 MW    0.00 MVAR
      7-  27 0   FDR71       0.55 MW        2.02 MVAR      2.09 MVA  LOSSES =    0.00 MW    0.00 MVAR
```

Fig 52
Printed Computer Output

system data, or the system base might be too large or too small for the computer to accurately represent the numbers involved. Short, low impedance lines in close proximity with long lines usually make convergence more difficult.

A load flow program requires at least one *swing* bus. This is a bus designated by the user as a bus on which system losses or excess power can be handled. Most programs also allow for a tie interchange or transfer of power from one system to another. Problems of converging to a solution can occur in some programs when zero impedance or high impedance lines are entered. Usually special handling is required for these lines and the methods can vary with different programs.

As mentioned briefly in 5.4.2, change cases can be run with only minor input changes on most programs. The usual procedure is to run a *base* case, usually an existing operating condition, which is checked against known conditions for accuracy. Then by changing, adding, or omitting lines, changing loads, and equipment, etc, change cases can be run consecutively as desired. In more sophisticated programs, provision is made for a data base where system parameters (equipment sizes, line impedances, bus numbers, loads, transformer sizes, voltages, ratios, taps, etc) are all stored for use in short circuit programs, load flow programs, and stability programs. The stored parameters provide data for the base case. Then changes are made by temporarily substituting the change data for the permanently stored data for a change case. Most programs using a data base allow for retaining any specified case, which can become the base case for future changes if desired.

5.7 Load Flow Analysis. A load flow study usually consists of several cases.

A case is defined as system power flow in all branches for given system data. It represents a balanced stable flow of power distributed throughout the system to satisfy Kirchhoff's laws and the law of conservation of energy. Current flow into a point is equal to that flowing out of the point (within an allowable small tolerance) and bus voltages and line drops are compatible throughout the system.

When analyzing a system load flow it should be noted that the real power (watt) is flowing in one direction from a given bus to another, while the imaginary power (var) is flowing in the opposite direction. This results from votlage on the second bus being greater than on the given bus. The real power must flow to the load, but to equalize the voltage, reactive power must flow in the opposite direction.

Proper transformer tap settings can reduce or change this condition. Many load flow programs have an automatic tap changing feature that changes taps to minimize reactive flow. Remember that var flow can be controlled by tap changing and that reactive power will flow from the bus with greater (relative) voltage. The taps must be selected to equalize this difference when done manually. Real power (watt) will flow to the load by the difference of the phase angle of the supply leading the phase angle of the load bus.

It must be stressed that input data to a load flow must be real values and as accurate as possible. Rounding off, or not including enough decimal places in certain parameters can be disastrous to the results in many cases. Do not ignore influential parameters. Results are no better than input.

The study of several such cases of a system, under the various operating conditions specified, leads to a knowledge of expected performance and behavior. Recognition of, and the appreciation for performance and behavior of the system under desired conditions is defined as a load flow analysis.

A load flow analysis is used to determine optimum bus voltages for normal operation, and yet continue to furnish reliable flows through alternate branches when one or more lines become inoperative due to line damage, lightning strokes, failure of transformers, etc. The study of multiple load flow cases and analysis of the results provide operating intelligence in a short time that might take years of actual operating experience to obtain.

In addition to optimum bus voltages, a study of reactive power flows in the branches can lead to reduced line losses, improved voltage distribution, and less var supply equipment. Transformer and line capacities are related to their maximum load flow requirements thus preventing burnout from overloads or adverse conditions. Transformer tap settings should be optimized to reduce reactive var flows to a minimum for practical operation. Knowledge of branch power flow supplies the protection engineer with requirements for proper relay settings to ensure normal operation and can provide data for automatic load and demand control if needed.

5.8 Load Flow Output Presentation. Maximum benefit results from a computer output report when power flows are graphically shown on a one-line diagram of the system. System flows can be quickly analyzed from this visual presentation which relates system configuration, operating conditions, and equipment parameters to an ideal or optimum operation. Another system one-line diagram is used to enter the case load flow results in much the same way the impedance diagram was prepared. Figure 53 illustrates this method using case 1 load flow results. In the illustration all pertinent system data for equipment and operating conditions are entered on the system one-line diagram. From the load flow report output, then bus voltage, voltage angle, load kW and kvar, capacitor kvar, etc are entered. By each interconnecting line (transmission line) the kW and kvar flowing in the line is entered. Note that arrows are drawn to show direction of real power. Reactive power then carries a positive (+) sign when it is flowing to the same bus as real power. When flowing in the opposite direction a minus (−) sign is shown. Where sizeable line loss occurs, power leaving the bus is shown near that bus, while power arriving at the receiving bus is shown near it with power loss indicated near the center of the line. At a convenient location on the drawing, or on a facing page in the analytical report, comments can be entered as to good or poor conditions, or both, that exist in circuit parameters or configuration. It is desirable to list corrective action taken for the next load flow run to hopefully improve the operation.

5.9 Load Flow Analysis. Now that a load flow has been run for conditions that exist, what has been learned about the system, and what can be done to improve the operation? Analysis of load flow output now presented on the one-line diagram shows the following:

(1) Voltage on bus 3 is low − 13.3, while voltage on bus 4 is 13.62

(2) Inspection of all bus voltages supplied from main bus 3 are relatively low

(3) Voltage on bus 39 is too low for good operation. This is due partially to low voltage on bus 3

(4) Generator 1 is absorbing vars when it should be supplying vars to the system

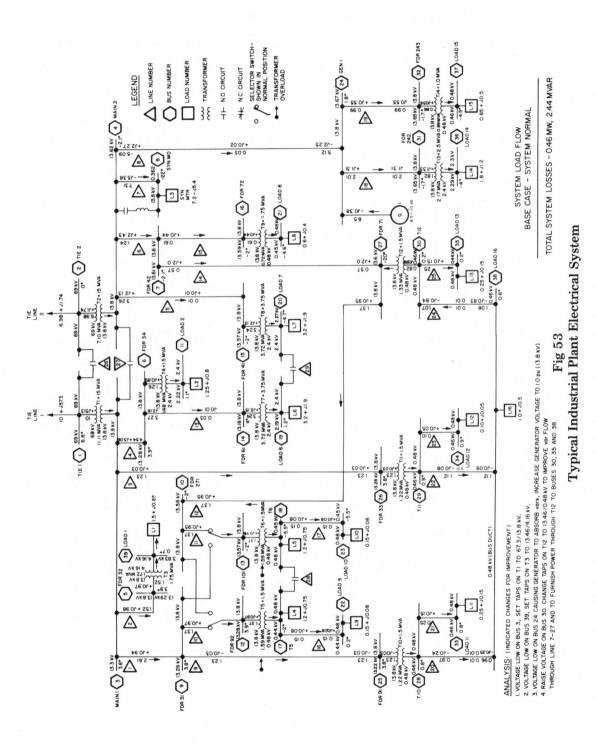

**Fig 53
Typical Industrial Plant Electrical System**

SYSTEM LOAD FLOW
BASE CASE – SYSTEM NORMAL

TOTAL SYSTEM LOSSES – 0.46 MW, 2.44 MVAR

ANALYSIS: (INDICATED CHANGES FOR IMPROVEMENT)
1. VOLTAGE LOW ON BUS 3, SET TAPS ON T1 TO 67.3/13.8 kV.
2. VOLTAGE LOW ON BUS 39, SET TAPS ON T3 TO 13.46/4.16 kV.
3. VOLTAGE LOW ON BUS 24 CAUSING GENERATOR TO ABSORB vars, INCREASE GENERATOR VOLTAGE TO 1.0 pu (13.8 kV).
4. RAISE VOLTAGE ON BUS 30. CHANGE TAPS ON T12 TO 13.46/0.48 kV TO IMPROVE var FLOW
   THROUGH LINE 7-27 AND TO FURNISH POWER THROUGH T12 TO BUSES 30, 35 AND 38.

(5) Var flow from bus 7 to bus 27 is greater than the real power flow

Buses that are being supplied with power from main bus 3 have lower than normal voltages. This suggests that taps should be changed on transformer T1 to raise the voltage on bus 3. This will increase the voltage on bus 5 and will help the low voltage condition on bus 39. However, changing taps on transformer T3 could raise bus voltage to a nominal 4.16 kV. Problem 4 can be remedied by increasing generator output voltage to a level high enough to supply about +j4 vars to the system. By changing the taps on transformer T12 to permit 0.48 kV on bus 30, the var flow through line 7–27 can be reduced while allowing it to reverse the power flow to bus 38 and to help supply power to load L16.

Such analysis of the system after each case run can tune the system gradually to obtain the most efficient and reliable operation.

Experience with load flows improves the engineer's ability to make corrections with a minimum number of case runs. However, it is stressed that any change affects the whole system, and a cure at one spot can create unexpected problems at another location in the system. For this reason, it is better not to make too many changes in a single run as the effects on the system may be difficult to understand. Each change case should be documented showing changes made and results obtained in order to keep future changes consistent with improving the system. A good place to record changes and their results is directly on the one-line diagram.

Some typical comments have been entered on Fig 53 under Analysis. Each analysis suggests what system changes can be made that would improve operation and reduce outages. New or larger transmission lines, new generation at optimum locations, new capacitors or reactors and power factor improvement are some items that analysis might show as desirable or necessary changes. Increasing loads over the years requires analysis to determine when, where, and how much new generation is required. Alternate generating schemes can be studied and analyzed. New transmission lines may be needed or possibly new tie lines to other utilities may be a solution.

5.10 Conclusions. It is evident to operators of industrial plant electrical systems, as well as utility system engineers, that a tool is needed to study their electrical systems under operating conditions that can actually be encountered *before* the condition occurs. Such a tool is used to prevent expensive outages, damaged equipment and possible loss of life. The load flow solution (usually as obtained from a relatively sophisticated computer program) is available and can be used to study systems under real or proposed conditions. The solution results should be evaluated and analyzed with respect to optimum present and future operation. This leads to a diagnosis of the system as it exists. The analysis can also point the way to improved operation and provide an adequate basis for future system planning.

6. Short-Circuit Studies

6.1 Introduction. Even the most carefully designed power systems are subject to damaging effects of high magnitude currents flowing from short circuits occurring in system components. To ensure that circuit protective equipment can isolate faults quickly and minimize damage and personnel hazard, it is essential that a short circuit analysis be included in the electrical design of new plants and also for modifications to existing plants.

A power system short circuit analysis can be used to determine any or all of the following:

(1) Calculated system fault current duties which can be compared with the first cycle (momentary) and interrupting short circuit current rating of circuit interrupting devices, such as circuit breakers and fuses

(2) Calculated system fault current duties to compare with short-time, or withstand ratings of system components such as cables, transformers, reactors, etc

(3) Selection and rating or setting of short circuit protective devices, such as direct-acting trips, fuses, and relays

(4) Evaluation of current flow and voltage levels in the overall system for short circuits in specific areas

Details of fundamental concepts involved in the rigorous calculation of short circuit currents are included in Section 3. Simplifying techniques and their limitations, and step-by-step procedures to follow for manual short circuit calculations are given in [1] and [2].

The intent of this section is to complement these sources by providing a brief overview of the steps required to perform a short circuit study and then illustrate how the computer can be used as an effective tool to aid in the calculations.

6.2 Short-Circuit Study Procedure

6.2.1 Preparing a One-Line Diagram. The starting point in performing a short-circuit study is the preparation of a basic system one-line diagram. Accuracy and usefulness of study results depends mainly on the reliability of this diagram. All major components such as motors, transformers, reactors, generators, utility supplies, feeder cables and ducts, and overhead lines should be shown. For an existing plant it is usually possible to obtain a one-line diagram from the engineering records. This diagram should be carefully checked to ensure that it has been updated to reflect accurately the system being studied. If such a diagram is not available, as may be the case for new (proposed) plants, it is necessary to prepare a suitable one-line diagram.

6.2.2 Determining Depth and Accuracy of a Study. The next step is to determine, from the basic one-line diagram, how extensive the short circuit study must be and what degree of accuracy is required.

There are five possibilities for a short circuit in a three-phase system:

(1) *Three-phase short circuit.* Three-phase conductors shorted together

(2) *Line-to-line short circuit.* Any two-phase conductors shorted together

(3) *Double line-to-ground short circuit.* Any two-phase conductors shorted together and simultaneously to ground

(4) *Single line-to-ground short circuit.* One-phase conductor shorted to ground

(5) *Three-phase to ground short circuit.*

For industrial power systems the most common study is the calculation of three-phase (balanced) short circuit current for comparison with switching equipment capability.

The short circuit current determined from this type of study generally represents the highest value at a particular location in the system. If information regarding lesser balanced short circuit values, or unbalanced short circuits (line-to-line, or line-to-ground) is not required, the basic study is all that is required.

It is important to realize that single line-to-ground short circuit current magnitude *can exceed* three-phase short circuit current under certain conditions. This can occur near (1) Solidly grounded synchronous machines, (2) Solidly

grounded wye (Y) connection of a delta-wye (Δ-Y) transformer of the three-phase core (three leg) design, (3) Grounded Y-Δ *tertiary* autotransformers, or (4) Grounded Y-grounded Y-Δ tertiary three winding transformers. In systems where these conditions exist, it is necessary to conduct a single line-to-ground short circuit study.

Note that in resistance grounded systems, unless it is essential to determine specific less-than-maximum, short circuit values, it is not necessary to calculate the ground short circuit current since neutral grounding resistors limit maximum short circuit current to a known value.

Careful consideration of the system one-line diagram indicates whether a three-phase short circuit study is adequate or whether one or more other conditions must also be studied.

6.2.3 Calculating Impedance Values. The next step is to determine values of impedance for all major passive and rotating (active) system components which are appropriate for the short circuit study.

The impedance of passive system elements is generally considered to be constant with respect to time (over the range of times considered in short circuit studies). For three-phase short circuit studies, only positive sequence impedance Z_1 is required. For phase-to-phase short circuit studies, positive and negative sequence impedances Z_1 and Z_2 are required. For single and double line-to-ground short circuit studies, positive, negative, and zero sequence impedance Z_1, Z_2, and Z_0 are required.

Each rotating (active) component is represented for short circuit purposes as a constant voltage source behind an impedance. This impedance varies with time (after short circuit) and so in addition to the different impedances listed

Table 8
Basic Impedance Values Required for
Three-Phase Short-Circuit Studies

	½ Cycle	Interrupting	Approximately 30 Cycle
Induction motors	X_d'', R	—	—
Synchronous motors and condensers	X_d'', R	X_d, R	—
Synchronous generators	X_d'', R	X_d'', R	X_d' or X_d
Electric utility systems	X_S, R_S	X_S, R_S	X_S, R_S

where

X_d' = transient reactance

X_d'' = subtransient reactance (for induction motors X_d'' is approximately equal to locked rotor reactance)

R = equivalent resistance

X_S, R_S = utility system equivalent reactance and resistance

Approximately 30 Cycle = a minimum source representation to check time delay relay sensitivity

for the passive elements of the various short-circuit studies, different impedance values are required for *½ cycle*, *interrupting*, and *approximately 30 cycle*, short-circuit studies.

Basic impedance values required for three-phase short-circuit studies are listed in Table 8.

In addition to impedance values listed for three-phase studies, machine negative sequence reactance X_2 is required for line-to-line studies and both X_2 and the zero sequence reactance X_0 are required for line-to-ground studies. Note that although positive sequence machine reactances (X_d'' and X_d') vary with time, negative and zero sequence reactances are constant.

Wherever possible, actual impedance values should be obtained for specific equipment. Many other excellent sources of typical data are available to assist when actual values are not known. The referenced sources [1], [2] contain such data.

6.2.4 Developing an Impedance Diagram. The next step is to modify the basic one-line diagram, or develop a new drawing from the one-line diagram, to include the impedances of all system components. This diagram is often called an impedance diagram.

6.2.5 Converting Impedances to a Common Base. The next step is to convert all impedances to a common base (this step can be done for the impedance diagram), so they can be combined in series and parallel and ultimately reduced to a single equivalent impedance to the point of short circuit.

Two established forms of expressing impedances are ohms and per unit (or percent which differs from per unit by a factor of 100). For short-circuit study purposes, the per unit (or percent) system is generally the simplest to use, particularly when the system under study has more than one voltage level.

Details on the development and use of the per unit system are given in 4.8.1.

6.2.6 Interpretation and Application of the Study. The final step involves calculation of short-circuit current, interpretation and application of results, and recommendations for system changes (in the case of existing plants) or for initial design (in the case of new plants).

General guidelines for interpretation and application of study results are usually not very useful since sound engineering judgment can only be made on the basis of treating each case individually. Some important questions to ask, however are:

(1) Is circuit interrupting equipment adequately rated for maximum short circuit momentary and interrupting availability? If not, what is the most economical method of making system changes while still maintaining a satisfactory degree of system flexibility?

(2) Is there any short-circuit capability margin for future expansion? If not, is it necessary? If it is necessary, what is the most suitable method of effecting changes to the system?

(3) Is non-interrupting equipment such as reactors, cables, interrupting equipment bus systems, bus duct, overhead lines, transformers, etc, adequately rated to withstand short-circuit current until cleared by circuit interrupting equipment?

(4) Is special protective equipment or circuitry necessary to provide protective device selectivity for both maximum and minimum values of short-circuit current?

(5) Does voltage of unfaulted buses in the system drop to values which will cause motor-starter contactor drop out or unnecessary operation of undervoltage relays? If so, is special equipment necessary to prevent a total system outage?

6.3 Use of the Computer. With the exception of system one-line diagram development and selection of impedance values for individual system components, computer short-circuit programs can be used in all other steps in the development of a system short-circuit study. Various programs are available which can be used for:

(1) Complex short-circuit studies using fundamental circuit analysis

(2) Short-circuit studies using simplifying techniques for *specific use* studies

(3) Only certain aspects of a study

Some of the most common uses ranging from simplest to most complex are:

(1) Reduction of a number of system impedances in series or parallel, or both, to a single equivalent impedance value

(2) Calculation of Y or Δ equivalent impedances to assist in impedance reduction

(3) Simple calculation of three-phase faults. No load flow data is required to establish initial conditions for input to the short circuit program. Output gives total fault E/X current, line currents to faulted buses and system voltages.

(4) Calculation of three-phase short circuit duties for use in comparing with interrupting device ratings. No load flow data required (see (3)). Output gives total fault E/X current, line current flows from other selected buses, X/R ratio values, and applies appropriate multipliers to fault values to allow direct comparison with interrupting device ratings based on applicable American National Standards.

(5) Calculations of three-phase, line-to-line, and line-to-ground short circuits in either simple or complex systems. Load flow data is required for input to establish initial conditions. Output gives positive, negative, and zero sequence line currents, and system voltages.

For most computer programs, user manuals are available for assistance in transferring information contained on the one-line diagram and the impedance diagram to a form which can be accepted by the computer. Program limitations and output format are generally also explained.

For some programs it is necessary to use cards as input data to the computer, while other programs, usually those used on a *time share* basis with small in-house terminals, use tape input and can be *conversational*. That is, the computer will ask questions which appear typed on the print-out sheet and the user then types out answers. This type of program is used in the solution of the short-circuit study example problem.

6.4 Short-Circuit Study Example. To illustrate the computer in calculating short circuit currents, an example problem will be studied. Basic steps previously described will be followed. The example problem involves a typical industrial plant situation where a three-phase short circuit study is necessary to ensure that switchgear and other equipment is adequately rated to withstand and interrupt the maximum short circuit current. Conditions of maximum load, maximum utility short circuit duty, and maximum generation are considered since this is the most severe fault condition.

The computer program used is on a *time share* basis and is *conversational* in mode. It has been specifically developed for calculating momentary (first ½ cycle) and interrupting short circuit values using techniques in accordance with ANSI/IEEE C37.13-1980 [3] for low voltage breakers and ANSI/IEEE C37.010-1979 [4] and ANSI/IEEE C37.5-1979 [5] for medium and high voltage breakers. ANSI/IEEE C37.010-1979 is used to compare short circuit values with circuit breakers rated on a symmetrical current basis (generally applicable for all new North American medium and high voltage breakers). ANSI/IEEE C37.5-1979 is used to compare short circuit values with circuit breakers rated on total current basis (generally applicable for older medium and high voltage North American breakers in existing plants).

Frequently interrupting current calculations are made using [6] as a guide. The principal extension this reference makes to American National Standards is a ratio of remote generator fault current to the sum of local generator fault current and remote generator fault current is used as a measure of electrical distance from the fault to the generation. The resulting fault current multiplier takes into account reactors and line impedances equivalent to transformer impedance, as well as variations in size of transformers.

The computer program in this example uses this interpretation. An interpolation between American National Standard Curves is included in the program, and the breaker duty, along with X/R ratio and breaker speed, is printed with remote/local-plus-remote interpolated multipliers applied to symmetrical current. The short circuit program prints enough information so other interpretations of the American National Standard can be used.

In the computer, the electrical power system is represented in matrix form. Each of the power system components (utility sources, generators, motors, transformers, cables, etc) is represented by a resistance value and a reactance value.

Note that in this example system X/R values are required, so both the reactance and the resistance of each component arc is required. In studies where X/R values are not required and when reactance of a component is very large compared to resistance, it is acceptable to neglect resistance; its value for computer input becomes zero.

The computer program places an assumed three-phase fault at the desired bus location in the system, and a set of short circuit currents is calculated for comparison with published short circuit ratings of the power system equipment.

Figure 54 is the one-line diagram for the system to be studied. It is the same diagram used in Section 5, Load Flow Example, and elsewhere in this text, with minor additions for specific use in short circuit study example.

Figure 55 is the impedance diagram for the system. It has been developed from Fig 54. All buses are numbered and all impedances have been put in a form which can be used as input to the computer. Major equipment impedances have been converted from values shown on the one-line diagram to per unit on a common 100 MVA base. Bus 38 load, which will not contribute short circuit current is not shown. Selected fault locations are shown as F 3, F 4, F 24, F 19, and F 30.

The utility system is represented by a per unit impedance which is equivalent to maximum short circuit duty (1000 MVA) available from the utility company.

Impedance for computer input is derived as follows:

$$X_{pu} = \frac{\text{base MVA}}{\text{utility fault availability}}$$

$$= \frac{100}{1000} = 0.1$$

$$R_{pu} = \frac{0.1}{22} = 0.0045$$

System cable impedances are typical values taken from Section 5 data.

Impedances for computer input are derived as follows:

$$X_{pu} \text{ or } R_{pu} = \frac{\text{actual ohms} \cdot \text{base MVA}}{(\text{base kV})^2}$$

Motors fed from each 480 V substation are grouped (lumped) and a single impedance is based on total connected motor kVA. This impedance was obtained by using hand calculations because the system is relatively small, but on a larger system, impedances can be obtained using a system data reduction computer program. Table 9 lists hand calculations and also using such a program, typical values of subtransient reactance X_d'' (locked rotor reactance X_{lr}) for each motor within the group are used and the total equivalent kVA and impedance is determined based on the following assumptions when exact motor impedances are not known:

Subtransient reactance (X_d'') values are used in first cycle (momentary) current calculations while a modified subtransient reactance is used in interrupting duties for medium and high voltage breakers. These values are in accordance with applicable circuit breaker application standards [4], [5].

American National Standards for calculating short circuit duties require actual motor or generator reactances be modified under certain conditions. Modification factors are listed in Table 10 for both momentary (close and latch) and interrupting duty calculations.

Table 10 shows a slight difference between momentary calculations using ANSI/IEEE C37.010-1979 and ANSI C37.5-1979 rather than using ANSI/IEEE C37.13-1980. In this example, the latter calculation is performed using low voltage calculations since the results are more conservative.

The X/R ratios of induction motors and transformers were determined by using *medium typical* curves from

114

Table 9
Assumed Values for Motors when
Exact Impedances are not Known

Induction motor	1 hp = 1 kVA
Synchronous motor, 0.8 PF	1 hp = 1 kVA
Synchronous motor, 1.0 PF	1 hp = 0.8 kVA
Lumped induction motors not greater than 600 V	$X_d'' = X_{lr} = 0.25$ per unit
Individual induction motors greater than 600 V	$X_d'' = X_{lr} = 0.17$ per unit
Synchronous motors not less than 1200 r/min	$X_d'' = 0.15$ per unit
Synchronous motors less than 1200 r/min but greater than 450 r/min	$X_d'' = 0.20$ per unit
Synchronous motors 450 r/min and less	$X_d'' = 0.28$ per unit

NOTE: Motor impedances are in per unit on motor kVA rating. Reactances and motor base kVA ratings listed above were taken from data and assumptions in [1].

Table 10
Modification Factors for Momentary and Interrupting Duty Calculations

Duty Calculation	System Component	Impedance Value for Medium and High Voltage Calculations per ANSI/IEEE C37.010-1979 and ANSI C37.5-1979	Impedance Value for Low Voltage Calculations ANSI/IEEE C37.13-1980*
First cycle (momentary) calculations	Utility supply	X_s	X_s
	Plant generators	X_d''	X_d''
	Synchronous motors	X_d''	X_d''
	Induction Motors		
	Above 1000 hp > 1200 r/min	X_d''***	X_d''***
	Above 250 hp > 1800 r/min	X_d''***	X_d''***
	All other motors		
	50–1000 hp	1.2 X_d''***	X_d''***
	Less than 50 hp	neglect	X_d''***
Interrupting calculations	Utility supply	X_s	**
	Plant generators	X_d''	**
	Synchronous motors	1.5 X_d''	**
	Induction Motors		
	Above 1000 hp > 1200 r/min	1.5 X_d''***	**
	Above 250 hp > 1800 r/min	1.5 X_d''***	**
	All other motors		
	50–1000 hp	3 X_d''***	**
	Less than 50 hp	neglect	

*Impedance (Z) values can be used for low voltage breaker duties.
**Not applicable.
***X_d'' for induction motors = locked rotor reactance.

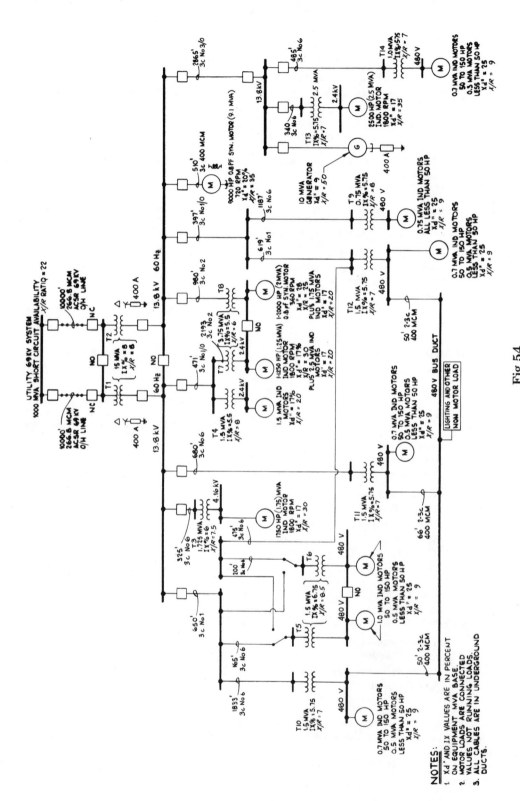

Fig 54
One-Line Diagram of Industrial System
for Short Circuit Study Example

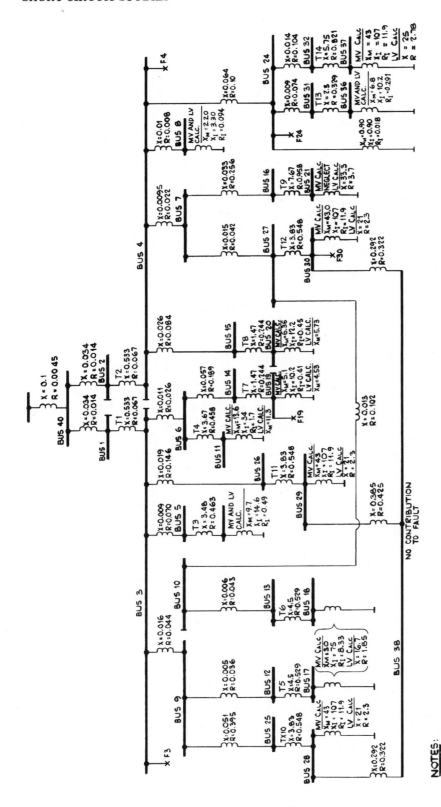

Fig 55
Impedance Diagram for Short-Circuit Study Example

NOTES:

1. ALL X & R VALUES ARE IN PER UNIT ON 100 MVA BASE

2. X_M = MOTOR REACTANCE FOR MOMENTARY CALCULATION

   X_I & R_I = MOTOR REACTANCE & RESISTANCE
   FOR INTERRUPTING CALCULATION

ANSI/IEEE C37.010-1979. When exact values can be obtained they should be used.

Examples of preparation of motor impedances for input to the computer are as follows:

(1) 13.8kV 9000 hp synchronous motors medium and low voltage calculations. See Fig 55, bus 8.

$$X_{momentary} = \frac{motor\ X_d'' \cdot base\ MVA}{motor\ rated\ MVA}$$
$$= \frac{0.20 \cdot 100}{9.1} = 2.20$$

$X_{interrupting} = 1.5\ (see\ Table\ 10) \cdot 2.20$
$= 3.30$

$$R_{interrupting} = \frac{3.30}{X/R} = \frac{3.30}{35} = 0.094$$

(2) 2.4 kV motors medium voltage calculation. See Fig 55, bus 20.
   (a) 2 MVA synchronous motor

$$X_{momentary} = 0.28 \cdot \frac{100}{2.0} = 14$$

$X_{interrupting} = 1.5 \cdot 14 = 21$

$$R_{interrupting} = \frac{21}{X/R} = \frac{21}{25} = 0.84$$

   (b) 1.75 MVA of induction motors between 50 hp and 1000 hp

$$X_{momentary\ lumped} = 0.17 \cdot \frac{100}{1.75}$$
$$\cdot 1.2 = 11.66$$

$$X_{interrupting\ lumped} = 0.17 \cdot \frac{100}{1.75} \cdot 3.0$$
$$= 29$$

$$R_{interrupting\ lumped} = \frac{29}{X/R} = \frac{29}{30}$$

   (c) Total combined motor impedances using the expression for combined X (or R)

$$X_{total} = \frac{X_1 X_2}{X_1 + X_2}$$
$$X_{momentary\ total} = \frac{14 \cdot 11.66}{14 + 11.66} = 6.36$$
$$X_{interrupting\ total} = \frac{21 \cdot 29}{21 + 29} = 12.2$$
$$R_{interrupting\ total} = \frac{0.84 \cdot 0.97}{0.83 + 0.97} = 0.45$$

(3) 2.4 kV motors low voltage calculation. See Fig 55, bus 20.
   (a) 2 MVA synchronous motor

$$X_{momentary} = 0.28 \cdot \frac{100}{2.0} = 14$$

   (b) 1.75 MVA of induction motors between 50 hp and 1000 hp

$$X_{momentary} = 0.17 \cdot \frac{100}{1.75} = 9.7$$

   (c) Total combined motor momentary impedance

$$X_{momentary\ total} = \frac{14 \cdot 9.7}{14 + 9.7} = 5.73$$

(4) 480 V motors medium voltage calculation. See Fig 55, bus 17.
   (a) 1.0 MVA of induction motors 50 to 150 hp

$$X_{momentary} = 0.25 \cdot \frac{100}{1.0} \cdot 1.2 = 30$$

$$X_{interrupting} = 0.25 \cdot \frac{100}{1.0} \cdot 3 = 75$$

$$R_{interrupting} = \frac{75}{X/R} = \frac{75}{9} = 8.33$$

   (b) 0.5 MVA of induction motors less than 50 hp neglect

(5) 480 V motors low voltage calculation. See Fig 55, bus 17.

$$X_{\text{momentary}} = \frac{0.25 \cdot 100}{\text{total motor MVA}}$$
$$= \frac{0.25 \cdot 100}{1.5}$$
$$= 16.7$$
$$R_{\text{momentary}} = \frac{16.7}{9} = 1.85$$

NOTE: Interrupting impedances are generally not applicable for low voltage calculations.

For the example all motors and the generator are assumed to be operating. This creates the highest possible short circuit currents the equipment may be subjected to since total short-circuit currents from all system motors, generator, and utility connection are present. Note, however, that the 13.8 kV bus tie is normally open and always will be unless one utility transformer is out of service. If the tie breaker were closed for normal operation the fault duty would be more severe and the study would be based on this operating mode.

The example study does not include prefault steady-state load currents. The effect of system load currents is usually negligible in short-circuit current studies for industrial and commercial power distribution systems.

6.5 Digital Computer Program Output Records

(1) Total and symmetrical short-circuit current duty at the faulted bus to compare directly with circuit breaker capability

(2) Short-circuit contributions from all buses connected to the faulted bus or between any other two buses specified in the input, or both

(3) Voltage at the remote buses where fault contributions are specified by (2)

(4) System X/R ratio at the fault point for medium voltage circuit breaker interrupting duty with its associated multiplying factor

(5) Symmetrical and asymmetrical current for breaker momentary duty

(6) Local and remote fault contributions.

To arrange system data contained on the impedance diagram so it can be accepted by the computer program, it is necessary to make up an input data tape for medium voltage momentary calculation, medium voltage interrupting calculation, and low voltage calculation. Figure 56 shows the data arranged for typing paper input data tape for medium voltage interrupting calculations.

Input data tape is a paper tape which, when fed into the computer, becomes a *file* for data storage. Change cases can then easily be run by modifying data lines in the file. The data file must be given a name. In this sample study the data file name for medium voltage interrupting calculation is CSP 100.

Figure 57 shows the program listing of input data from data file CSP 100. Data is usually listed in this way so it can be checked for errors before proceeding with short circuit calculations.

Figure 58 is a sample of the computer output for the medium voltage interrupting case giving remote bus voltages in per unit of normal voltage, and short-circuit contributions in MVA.

Figure 59 is computer output showing total fault level in MVA at each faulted bus, and also contributions from all connecting buses.

Also shown are X/R ratios at the faulted bus, multipliers taken from the standards to apply to the E/X values for both 8 cycle and 5 cycle breakers, and the fault duty for direct comparison with the circuit breaker rating. Remote (in this case the utility system), and local (in this case the in-plant generator), sources of short

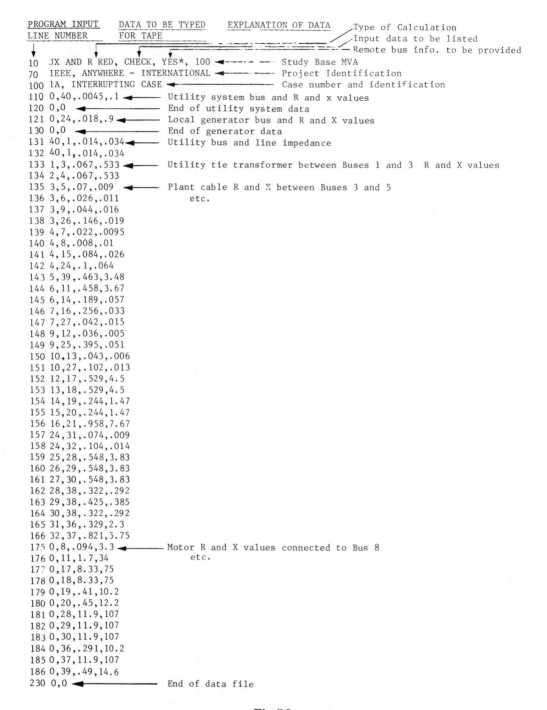

```
PROGRAM INPUT      DATA TO BE TYPED    EXPLANATION OF DATA        Type of Calculation
LINE NUMBER        FOR TAPE                                       Input data to be listed
                                                                  Remote bus info. to be provided
 10  JX AND R RED, CHECK, YES*, 100 ◄----- --- Study Base MVA
 70  IEEE, ANYWHERE - INTERNATIONAL ◄------- ---- Project Identification
100  1A, INTERRUPTING CASE ◄─────────────── Case number and identification
110  0,40,.0045,.1 ◄──────── Utility system bus and R and x values
120  0,0 ◄──────────── End of utility system data
121  0,24,.018,.9 ◄─────── Local generator bus and R and X values
130  0,0 ◄────────── End of generator data
131  40,1,.014,.034 ◄────── Utility bus and line impedance
132  40,1,.014,.034
133  1,3,.067,.533 ◄─────── Utility tie transformer between Buses 1 and 3  R and X values
134  2,4,.067,.533
135  3,5,.07,.009 ◄──────── Plant cable R and X between Buses 3 and 5
136  3,6,.026,.011             etc.
137  3,9,.044,.016
138  3,26,.146,.019
139  4,7,.022,.0095
140  4,8,.008,.01
141  4,15,.084,.026
142  4,24,.1,.064
143  5,39,.463,3.48
144  6,11,.458,3.67
145  6,14,.189,.057
146  7,16,.256,.033
147  7,27,.042,.015
148  9,12,.036,.005
149  9,25,.395,.051
150  10,13,.043,.006
151  10,27,.102,.013
152  12,17,.529,4.5
153  13,18,.529,4.5
154  14,19,.244,1.47
155  15,20,.244,1.47
156  16,21,.958,7.67
157  24,31,.074,.009
158  24,32,.104,.014
159  25,28,.548,3.83
160  26,29,.548,3.83
161  27,30,.548,3.83
162  28,38,.322,.292
163  29,38,.425,.385
164  30,38,.322,.292
165  31,36,.329,2.3
166  32,37,.821,5.75
175  0,8,.094,3.3 ◄──────── Motor R and X values connected to Bus 8
176  0,11,1.7,34               etc.
177  0,17,8.33,75
178  0,18,8.33,75
179  0,19,.41,10.2
180  0,20,.45,12.2
181  0,28,11.9,107
182  0,29,11.9,107
183  0,30,11.9,107
184  0,36,.291,10.2
185  0,37,11.9,107
186  0,39,.49,14.6
230  0,0 ◄──────── End of data file
```

Fig 56
Data Taken from the Impedance Diagram and Arranged for
Program Input Data Paper Tape Medium Voltage Interrupting Calculation

NAME OF DATA FILE? csp100 ◄———————— NOTE: USER MUST TYPE IN ANSWERS
INPUT DATA   TO QUESTIONS

BUS	TO	BUS	=	R	JX
0		40		0.0045	0.1
0		24		0.018	0.9
1		40		0.014	0.034
2		40		0.014	0.034
1		3		0.067	0.533
2		4		0.067	0.533
3		5		0.07	0.009
3		6		0.026	0.011
3		9		0.044	0.016
3		26		0.146	0.019
4		7		0.022	0.0095
4		8		0.008	0.01
4		15		0.084	0.026
4		24		0.1	0.064
5		39		0.463	3.48
6		11		0.458	3.67
6		14		0.189	0.057
7		16		0.256	0.033
7		27		0.042	0.015
9		12		0.036	0.005
9		25		0.395	0.051
10		13		0.043	0.006
10		27		0.102	0.013
12		17		0.529	4.5
13		18		0.529	4.5
14		19		0.244	1.47
15		20		0.244	1.47
16		21		0.958	7.67
24		31		0.074	0.009
24		32		0.104	0.014
25		28		0.548	3.83
26		29		0.548	3.83
27		30		0.548	3.83
28		38		0.322	0.292
29		38		0.425	0.385
30		38		0.322	0.292
31		36		0.329	2.3
32		37		0.821	5.75
0		8		0.094	3.3
0		11		1.7	34
0		17		8.33	75
0		18		8.33	75
0		19		0.41	10.2
0		20		0.45	12.2
0		28		11.9	107
0		29		11.9	107
0		30		11.9	107
0		36		0.291	10.2
0		37		11.9	107
0		39		0.49	14.6

SYSTEM HAS 1 LOCAL SOURCES AND 1 REMOTE SOURCES ◄——— LOCAL SOURCE = GENERATOR
                                                     REMOTE SOURCE = UTILITY

INPUT DATA CORRECT? yes
INPUT DATA LISTED IN OUTPUT? yes

PRINT ALL BUSES? no
 3 CYC. MF FOR WHICH 5 BUSES? 3,4,24,19,0 ◄——————— ONLY SELECTED FAULTED
PRINT ALL BUSES? no   BUSES ARE CONSIDERED IN
 3 CYC. MF FOR WHICH 5 BUSES? 0,0,0,0,0   THIS EXAMPLE STUDY
PRINT ALL BUSES? NO
INPUT 5 BUS NUMBERS? 3,4,24,19,0

Fig 57
Program Listing of Input Data from Data Tape

```
REMOTE MVA CONTRIBUTIONS-CASE 1A          NOTE: USER MUST TYPE IN ANSWERS
                                                TO QUESTIONS
BUS # FAULTED    ? 3
  REMOTE LINE  ? 40,1
  CONTRIBUTION FROM  40 ( 0.854 V) TO   1 ( 0.803 V)= 150.615  MVA

  REMOTE LINE  ? 40,2
  CONTRIBUTION FROM  40 ( 0.854 V) TO   2 ( 0.856 V)= 4.59  MVA

  REMOTE LINE  ? 2,4
  CONTRIBUTION FROM   2 ( 0.856 V) TO   4 ( 0.88 V)= 4.59  MVA

  REMOTE LINE  ? 4,15
  CONTRIBUTION FROM   4 ( 0.88 V) TO  15 ( 0.88 V)= 0.876  MVA

  REMOTE LINE  ? 4,7
  CONTRIBUTION FROM   4 ( 0.88 V) TO   7 ( 0.879 V)= 13.283  MVA

  REMOTE LINE  ? 4,8
  CONTRIBUTION FROM   4 ( 0.88 V) TO   8 ( 0.88 V)= 3.627  MVA

  REMOTE LINE  ? 4,24
  CONTRIBUTION FROM   4 ( 0.88 V) TO  24 ( 0.889 V)= 13.372  MVA

  REMOTE LINE  ? 10,27
  CONTRIBUTION FROM  10 ( 0.877 V) TO  27 ( 0.877 V)= 0.154  MVA

  REMOTE LINE  ? 0,0

BUS # FAULTED    ? 4
  REMOTE LINE  ? 40,1
  CONTRIBUTION FROM  40 ( 0.843 V) TO   1 ( 0.84 V)= 8.233  MVA

  REMOTE LINE  ? 40,2
  CONTRIBUTION FROM  40 ( 0.843 V) TO   2 ( 0.793 V)= 148.691  MVA

  REMOTE LINE  ? 1,3
  CONTRIBUTION FROM   1 ( 0.84 V) TO   3 ( 0.796 V)= 8.233  MVA

  REMOTE LINE  ? 3,5
  CONTRIBUTION FROM   3 ( 0.796 V) TO   5 ( 0.796 V)= 1.127  MVA

  REMOTE LINE  ? 3,6
  CONTRIBUTION FROM   3 ( 0.796 V) TO   6 ( 0.797 V)= 2.275  MVA

  REMOTE LINE  ? 3,9
  CONTRIBUTION FROM   3 ( 0.796 V) TO   9 ( 0.795 V)= 5.724  MVA

  REMOTE LINE  ? 3,26
  CONTRIBUTION FROM   3 ( 0.796 V) TO  26 ( 0.795 V)= 5.908  MVA

  REMOTE LINE  ? 10,27
  CONTRIBUTION FROM  10 ( 0.004 V) TO  27 ( 0.004 V)= 1.248  MVA

  REMOTE LINE  ? 0,0
```

Fig 58
Sample Computer Output Listing of Remote Bus Voltages
and Short Circuit Contributions to the Faulted Bus
Medium Voltage Interrupting Case Short Circuit Study.

RESULTS IN MVA

BUS   3
E/X= 183.915 MVA   X/R= 8.209
MF FOR 8 CY(.0667 SEC. CONT. PART)TOT CB = 1 ,CB DUTY = 183.915
MF FOR 5 CY(.0500 SEC. CONT. PART)SYM CB = 1 ,CB DUTY = 183.915
CONTRIBUTION :
BUS 1 = 150.613   BUS 5 = 5.5175   BUS 6 = 11.1686
BUS 9 = 8.9544   BUS 26 = 7.6373

    BUS NO.   LOCAL(MVA)   REMOTE(MVA)   GEN VOLTS
      40   0   146.023   0.854
      24   1.38   11.002   0.889
      REMOTE / REMOTE+LOCAL = 0.991

BUS   4
E/X= 312.053 MVA   X/R= 10.262
MF FOR 8 CY(.0667 SEC. CONT. PART)TOT CB = 1 ,CB DUTY = 312.053
MF FOR 5 CY(.0500 SEC. CONT. PART)SYM CB = 1 ,CB DUTY = 312.053
CONTRIBUTION :
BUS 2 = 148.69   BUS 7 = 14.4201   BUS 8 = 30.2067
BUS 15 = 7.2972   BUS 24 = 111.437

    BUS NO.   LOCAL(MVA)   REMOTE(MVA)   GEN VOLTS
      40   0   156.924   0.843
      24   95.827   7.359   0.071
      REMOTE / REMOTE+LOCAL = 0.632

BUS 24
E/X= 297.784 MVA   X/R= 21.523
MF FOR 8 CY(.0667 SEC. CONT. PART)TOT CB = 1.048 ,CB DUTY = 312.087
MF FOR 5 CY(.0500 SEC. CONT. PART)SYM CB = 1.018 ,CB DUTY = 303.103
CONTRIBUTION :
BUS 4 = 177.791   BUS 31 = 7.974   BUS 32 = 0.8721
GEN = 111.111

    BUS NO.   LOCAL(MVA)   REMOTE(MVA)   GEN VOLTS
      40   0   139.068   0.861
      24   111.111   0   0
      REMOTE / REMOTE+LOCAL = 0.556

BUS 19
E/X= 57.2469 MVA   X/R= 7.565
MF FOR 8 CY(.0667 SEC. CONT. PART)TOT CB = 1 ,CB DUTY = 57.247
MF FOR 5 CY(.0500 SEC. CONT. PART)SYM CB = 1 ,CB DUTY = 57.247
CONTRIBUTION :
BUS 14 = 47.4432   GEN = 9.8039

    BUS NO.   LOCAL(MVA)   REMOTE(MVA)   GEN VOLTS
      40   0   39.485   0.961
      24   0.101   3.247   0.97
      REMOTE / REMOTE+LOCAL = 0.998

Fig 59
Computer Output Giving Fault Levels in MVA for
the Four Faulted Buses, Medium Voltage Interrupting Case Short Circuit Study

Table 11
Sample Summary of Results for Example Short Circuit Study

| Short Circuit Location | Short Circuit Duty to Compare Directly with Equipment Rating | | Voltage in Per Unit on Associated Buses (Interrupting Case Calculations) | | | |
	Momentary (first ½ cycle) Asym kA	Interrupting 5 Cycle Sym MVA	Bus 1	Bus 4	Bus 3	Bus 24
13.8 kV Bus 3 (F 3)	13.4	183.9	0.80	0.88	0	0.89
13.8 kV Bus 4 (F 4)	22.7	312	0.84	.0	0.80	.0 +
13.8 kV Bus 24 (F 24)	21.4	303	0.86	0.11	0.82	.0
2.4 kV Bus 19 (F 19)	26.1	57.2	0.95	0.97	0.74	0.97
480 V Bus 30 (F 30)	68.9*	Not applicable	0.96	0.92	0.92	0.92

*Low voltage calculation in first ½ cycle symmetrical kA.

circuit current are listed separately.

Table 11 shows a sample summary of the short-circuit study results. Not all computer output information is shown since it is not possible to clearly present this information on one table. Several tables might be necessary to present complete study results, or in some cases, short circuit diagrams showing all system voltages and short circuit flows, provide the best means of presenting study results.

Using the complete computer output information, study results can be analyzed and applied to specific equipment shown on the one-line diagram.

6.5 References. The following references were used in the preparation of Section 6.

[1] IEEE Std 141-1976, IEEE Recommended Practice for Electric Power Distribution for Industrial Plants.

[2] IEEE Std 242-1975, IEEE Recommended Practice for Protection and Coordination of Industrial and Commercial Power Systems.

[3] ANSI/IEEE C37.13-1980, IEEE Standard for Low Voltage AC Power Circuit Breakers Used in Enclosures.

[4] ANSI/IEEE C37.010-1979, IEEE Application Guide for AC High Voltage Circuit Breakers Rated on a Symmetrical Current Basis (Consolidated edition).

[5] ANSI/IEEE C37.5-1979, IEEE Guide for Calculation of Fault Currents for Application of AC High-Voltage Circuit Breakers Rated on a Total Current Basis (Revision of ANSI C37.5-1969).

[6] IEEE Transactions Paper 69TP146-IGA Sep/Oct 1969, Interpretation of New American National Standards for Power Circuit Breaker Application.

7. Transient Stability Studies

7.1 Introduction. For years, system stability has been a problem almost exclusively to electric utility engineers. Within the past decade, however, increasing numbers of industrial and commercial facilities have installed local generation, large synchronous motors, or both. This means that system stability is of concern to a growing number of industrial plant electrical engineers and consultants.

7.2 Stability Fundamentals

7.2.1 Definition of Stability. Fundamentally, stability is a property of a power system containing two or more synchronous machines. The system is stable, *under a specified set of conditions*, if all of its synchronous machines remain in step with one another (or having pulled out of step, regain synchronism soon afterwards). The emphasis on specified conditions in this definition is intended to stress the fact that a system which is stable under one set of conditions can be unstable under some other set of conditions.

7.2.2 Steady-State Stability. Although the discussion in the rest of this section revolves around stability under transient conditions such as faults, switching operations, etc, there should also be an awareness that a power system can become unstable under steady-state conditions.

The simplest power system to which stability considerations apply consists of a pair of synchronous machines, one acting as a generator and the other acting as a motor, connected together through a reactance. See Fig 60. (In this model the reactance is the sum of the transient reactances of the two machines and the

Fig 60
Simplified Two-Machine
Power System

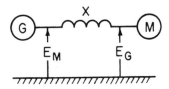

reactance of the connecting circuit. Losses in the machines and the resistance of the line are neglected for simplicity).

If the internal voltages of the two machines are E_G and E_M, and the phase angle between them is θ, it can easily be demonstrated [1], [2] that the real power transmitted from the generator to the motor is

$$P = \frac{E_G \, E_M}{X} \, \sin \theta$$

The maximum value of P obviously occurs when $\theta = 90°$. Thus

$$P_{max} = \frac{E_G \, E_M}{X}$$

This is the steady-state stability limit for the simplified two-machine system. Any attempt to transmit more power than P_{max} will cause the two machines to pull out of step, with the given values of internal voltages.

This model of a simple power system shows that at least three electrical characteristics of the system affect stability. They are:

(1) Internal voltage of the generator

(2) Reactances of the machines and transmission system

(3) Internal voltage of the motor

The higher the internal voltages, and the lower the system and machine reactances, the greater power can be transmitted under steady-state conditions.

7.2.3 Transient Stability. The preceding look at steady-state stability serves as a background for an examination of the more complicated problem of transient stability. This is true because the same three electrical characteristics which determine steady-state stability limits, affect transient stability in the same way. However, a system which is stable under

steady-state conditions is not necessarily stable when subjected to a transient disturbance.

Transient stability means the ability of a power system to survive a sudden change in generation, load, or system characteristics without a prolonged loss of synchronism. To see how a disturbance affects a synchronous machine, first look at the steady-state characteristics described by the steady-state torque equation [3].

$$T = \frac{\pi P^2}{8} \, \phi_{SR} \, F_R \, \sin \delta_R$$

where

T = mechanical shaft torque

P = number of poles of machine

ϕ_{SR} = air-gap flux

F_R = rotor field MMF

δ_R = mechanical angle between rotor and stator field lobes

The air-gap flux ϕ_{SR} stays constant as long as the voltage at the machine does not change, if the effects of saturation of the iron are neglected. Therefore, if the field excitation remains unchanged, a change in shaft torque T will cause a corresponding change in rotor angle δ_R. (This is the angle by which, for a motor, the peaks of the rotating stator field lead the corresponding peaks of the rotor field. For a generator, the relation is reversed.) Fig 61 graphically illustrates the variation of rotor angle with shaft torque. With the machine operating as a motor (when rotor angle and torque are positive), torque increases with rotor angle until δ_R reaches 90 electrical degrees. Beyond 90°, torque decreases with increasing rotor angle. As a result, if we increase the required torque output of a synchronous motor beyond the level corresponding to 90° rotor angle, it will *slip*

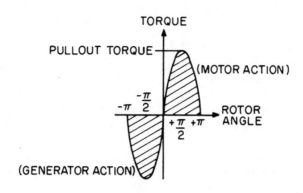

Fig 61
Torque Versus Rotor Angle Relationship for
Synchronous Machines in Steady State

a pole. Unless the load torque is reduced below the 90° level (the pullout torque), the motor will continue slipping poles indefinitely. The problems that can follow from extended operation in this out-of-step condition will be discussed further in this section.

A generator operates similarly. Increasing torque input until the rotor angle exceeds 90° results in pole slipping and loss of synchronism with the power system, assuming constant electrical load.

Similar relations apply to the other parameters of the torque equation. For example, air-gap flux ϕ_{SR} is a function of voltage at the machine. Thus if the other factors remain constant, a change in system voltage will cause a change in rotor angle. Likewise, changing the field excitation will cause a change in rotor angle, if constant torque and voltage are maintained.

The preceding discussion refers to rather gradual changes in the conditions affecting the torque angle, so that approximately steady-state conditions always exist. The coupling between the stator and rotor fields of a synchronous machine, however, is somewhat elastic. This means that if an abrupt rather than

a gradual change occurs in one or more of the parameters of the torque equation, the rotor angle will tend to overshoot the final value determined by the changed conditions. This disturbance can be severe enough to carry the ultimate steady-state rotor angle past 90°, or the transient *swing* rotor angle past 180°. Either event results in the slipping of a pole. If the conditions which caused the original disturbance are not corrected, the machine will then continue to slip poles, in short, pull out of step with the power system to which it is connected.

Of course, if the rotor angle overshoot does not transitorily exceed 180°, or if the disturbance causing the rotor swing is promptly removed, the machine may remain in synchronism with the system. The rotor angle then oscillates in decreasing swings until it settles to its final value (less than 90°). The oscillations are damped by mechanical load and losses in the system, especially in the damper windings of the machine.

A change in rotor angle of a machine generally requires a change in speed of the rotor. For example, if we assume that the stator field frequency is constant, it is necessary to at least momentarily slow down the rotor of a synchronous motor to permit the rotor field to fall farther behind the stator field and thus increases δ_R. The rate at which rotor speed can change is determined by the moment of inertia of the rotor, plus whatever is mechanically coupled to it, prime mover, load, reduction gears, etc. This means a machine with high inertia is less likely to pull out under a disturbance of brief duration than a low-inertia machine, all other characteristics being equal.

7.2.4 Two-Machine Systems. The previous discussion of transient behavior of synchronous machines is based on a single machine connected to a good approximation of an infinite bus. An example is the typical industrial situation where a synchronous motor of at most a few thousand horsepower, is connected to a utility company system with a capacity of thousands of megawatts. Under these conditions we can safely neglect the effect of the machine on the power system.

A system consisting of only two machines of comparable size connected through a transmission link, however, becomes more complicated, because the two machines can affect one another's performance. The medium through which this occurs is the air gap flux. This is a function of machine terminal voltage, which is affected by the characteristics of the transmission system, the amount of power being transmitted, and the power factor, etc.

In the steady state, the rotor angles of the two machines are determined by the simultaneous solution of their respective torque equations. Under a transient disturbance, as in the one-machine system, the rotor angles move toward values corresponding to the changed system conditions. Even if these new values are within the steady-state stability limits of the system, an overshoot can result in loss of synchronism. If not, both rotors will undergo a damped oscillation and ultimately settle to their new steady-state values.

An important concept here is synchronizing power. The more real power transmitted over the transmission link between the two machines, the more likely they are to remain in synchronism in the face of a transient disturbance. Synchronous machines separated by a sufficiently low impedance behave as one composite machine, since they tend to remain in step with one another regardless of external disturbances.

7.2.5 Multimachine Systems. At first glance, it appears that a power system incorporating many synchronous machines would be extremely complex to analyze. This is true if a detailed, precise analysis is needed; a large digital computer and a sophisticated program are required for a complete transient stability study of a multimachine system. However, many of the multimachine systems encountered in industrial practice contain only synchronous motors which are similar in characteristics, closely coupled electrically, and connected to a high-capacity utility system. Under most types of disturbance, motors will remain synchronous with each other, although they can all lose synchronism with the utility. Thus, the problem is similar to a single synchronous motor connected through an impedance to an infinite bus. The simplification should be apparent. More complex systems, where machines are of comparable sizes and are separated by substantial impedances, will usually involve a full-scale computer stability study.

7.3 Problems Caused by Instability. The most immediate hazards of asynchronous operation of a power system are the high transient mechanical torques and currents which usually occur. To prevent these transients from causing mechanical and thermal damage, synchronous motors and generators are almost universally equipped with pullout protection. For motors of small to moderate sizes, this protection is usually provided by a damper protection of pullout relay which operates on the low power factor occurring during asynchronous operation. The same function is usually provided for large motors, generators, and synchronous condensers by loss-of-field relaying. In any case, the pullout relay trips the machine breaker or contactor. Whatever load is being served by the machine is naturally interrupted. Consequently, the primary disadvantage of a system which tends to be unstable is the probability of frequent process interruptions.

Out-of-step operation also causes large oscillatory flows of real and reactive power over the circuits connecting the out-of-step machines. Impedance or distance-type relaying protecting these lines can falsely interpret power surges as a line fault, tripping the line breakers and breaking up the system. Although this is primarily a utility problem, large industrial systems or those where local generation operates in parallel with the utility can be susceptible.

7.4 System Disturbances that Can Cause Instability. The most common disturbances that produce instability in industrial power systems are (not necessarily in order of probability):

(1) Short circuits

(2) Loss of a tie circuit to a public utility

(3) Loss of a portion of on-site generation

(4) Starting a motor which is large relative to a system generating capacity

(5) Switching operations

(6) Impact loading on motors

(7) Abrupt decrease in electrical load on generators

The effect of each of these disturbances should be apparent from the previous discussion of stability fundamentals. Items (1) through (5) tend to reduce voltage levels, ultimately requiring an increase in machine angles to maintain a given load. Items (6) and (7) directly increase the rotor angles of affected machines.

7.5 Solutions to Stability Problems. Generally, anything which decreases the severity or duration of a transient disturbance will make the power system less likely to become unstable under that disturbance. In addition, increasing the moment of inertia per rated kVA of the synchronous machines in the system will raise stability limits by resisting changes in rotor speeds required to change rotor angles.

7.5.1 System Design. System design primarily affects the amount of synchronizing power that can be transferred between machines. Two machines connected by a low-impedance circuit such as a short cable or bus run will probably stay synchronized with each other under all conditions except a fault on the connecting circuit, a loss of field excitation, or an overload. The greater the impedance between machines, the less severe a disturbance will be required to drive them out of step. This means that from the standpoint of maximum stability all synchronous machines should be closely connected to a common bus. Limitations on short circuit duties, economics, and the requirements of physical plant layout usually combine to render this radical solution impractical.

7.5.2 Design and Selection of Rotating Equipment. Design and selection of rotating equipment can be a major contributor to improving system stability. Most obviously, use of induction instead of synchronous motors eliminates the potential stability problems associated with the latter. (Under rare circumstances an induction motor/synchronous generator system can experience instability, in the sense that undamped rotor oscillations occur in both machines, but the possibility is too remote to be of serious concern.) However, economic considerations often preclude this solution.

Where synchronous machines are used, stability can be enhanced by increasing the inertia of the mechanical system. Since the H constant (stored energy per rated kVA) is proportional to the square of the speed, fairly small increases in synchronous speed can pay significant dividends in higher inertia. If carried too far, this can become self-defeating, because higher speed machines have smaller diameter rotors. WK^2 varies with the square of the rotor radius, so the increase in H due to higher speed may be offset by a decrease due to the lower WK^2 of a smaller diameter rotor.

A further possibility is to use synchronous machines with low transient reactances that permit the maximum flow of synchronizing power. Applicability of this solution is limited mostly by short circuit considerations and machine design problems.

7.5.3 System Protection. System protection often offers the best prospects for improving the stability of a power system. The most severe disturbance which an industrial power system is likely to experience is a short circuit. To prevent loss of synchronism, as well as to limit personnel hazards and equipment damage, short circuits should be isolated

as rapidly as possible. A system which tends to be unstable should be equipped with instantaneous overcurrent protection on all of its primary feeders, which are the most exposed section of the primary system. As a general rule, instantaneous relaying should be used throughout the system wherever selectivity permits.

7.5.4 Voltage Regulator and Exciter Characteristics. Voltage regulator and exciter characteristics affect stability because, all other things being equal, higher field excitation requires a smaller rotor angle. Consequently, stability is enhanced by a properly applied regulator and exciter which respond rapidly to transient effects and furnish a high degree of field forcing. In this respect, modern solid-state voltage regulators and static exciters can contribute markedly to improved stability. (On the other hand, a mismatch in exciter and regulator characteristics can make an existing stability problem even worse.)

7.6 Transient Stability Studies. Knowing how to correct an unstable power system is not very valuable if, in order to test our proposed recommendations, we have to stage stability tests on the actual system. This is especially true if the system in question is still in the design stage. Consequently, we need a fast, simple, and inexpensive way to simulate transient performance of a power system under a variety of normal and abnormal conditions.

7.6.1 History. The first transient stability studies responding to this need were done by semieducated guesses. When this technique proved insufficiently accurate, and as rotating-machine theory was developed, simple power systems consisting of only a few machines were analyzed by manual calculations or mechanical models. These methods were neither precise nor applicable to complex systems. In the thirties and forties, further development of modeling techniques led to the use of the ac network analyzer in stability studies. This analog device permitted the simulation of much larger systems than previously possible, but still suffered from the disadvantages of imprecise representation of synchronous machines, proneness to human error, limited capacity, and high cost per study run. Finally in the fifties the digital computer came on the scene. Its enormous arithmetic capability, precision, and ability to store and retrieve huge amounts of information made it a natural for stability studies. Transient stability programs were written and used by major electrical suppliers, utility companies, consulting firms, and universities, and have now almost totally replaced the older methods.

7.6.2 How Stability Programs Work. Mathematical methods of stability analysis depend on a repeated solution of the swing equation for each machine,

$$P_a = \frac{\text{MVA } (H)}{180\,f}\,\frac{d^2\delta_R}{dt^2}$$

where

P_a = accelerating power (input power minus output power), MW

MVA = rated MVA of machine

H = inertia constant of machine, MW·s/MVA

f = system frequency, Hz

δ_R = rotor angle, degrees

t = time, s

The program begins with the results of a load-flow study to establish initial power and voltage levels in all machines and interconnecting circuits. The specified disturbance is applied at a time defined as zero, and the resulting changes in power levels are calculated by a load flow routine. Using the calculated accelerating power values, the swing equation is solved for a new value of δ_R for each machine at an incremental time (for example, 0.01 s) after the disturbance. Voltage and power levels corresponding to this new angular position of the synchronous machines are then used as base information for another iteration. In this way, performance of the system is calculated for every integration interval out to as much as several minutes.

7.6.3 Simulation of the System. A modern transient stability study computer program can simulate virtually any set of power system components in sufficient detail to give accurate results. Simulation of rotating machines and related equipment is of special importance in stability studies. The simplest possible representation for a synchronous motor or generator involves only a constant internal voltage, a constant transient reactance, and the rotating inertia (H) constant. This approximation neglects saturation of core iron, voltage regulator action, the influence of construction of the machine on transient reactances for the direct and quadrature axes, and most of the characteristics of the prime mover or load. Nevertheless, this so-called *classical* representation is often accurate enough to give reliable results, especially when the time period being studied is rather short. (Limiting the study to a short period — say, ½ second or less, means that neither the voltage regulator nor the governor, if any, has time to exert a significant effect.) The classical representation is generally used for the smaller and less influential machines in a system, or where the more detailed information required for better simulations is not available.

As additional data on the machines become available, better approximations can be used. This permits more accurate results which remain reliable for longer time periods. Modern large-scale stability programs can simulate all of the following characteristics of a rotating machine:

(1) Voltage regulator and exciter
(2) Steam system or other prime mover, including governor
(3) Mechanical load
(4) Damper windings
(5) Salient poles
(6) Saturation

Induction motors can also be simulated in detail, together with speed-torque characteristics of their connected loads.

In addition to rotating equipment, the stability program can include in its simulation practically any other major system component, including transmission lines, transformers, capacitor banks, and voltage regulating transformers and dc transmission links in some cases.

7.6.4 Simulation of Disturbances. The versatility of the modern stability study is apparent in the range of system disturbances that can be represented. The most severe disturbance which can occur on a power system is usually a three-phase bolted short circuit. Consequently, this type of fault is most often used to test system stability. Stability programs can simulate a three-phase fault at any location, with provisions for clearing the fault by opening breakers either after a specified time delay, or by the action of overcurrent, underfrequency, overpower, or impedance relays. This feature permits the adequacy of proposed protective relaying to be evaluated from the stability standpoint.

Short circuits other than the bolted three-phase fault cause less disturbance to the power system. Although most stability programs cannot directly simulate line to line or ground faults, the effects of these faults on synchronizing power flow can be duplicated by applying a three-phase fault with a properly chosen fault impedance. This means the effects of any type of fault on stability can be studied.

In addition to faults, stability programs can simulate switching of lines and generators. This is particularly valuable in the load-shedding type of study, which will be covered in the following section.

Finally, starting of large motors on relatively weak power systems and impact loading of running machines can be analyzed.

7.6.5 Data Requirements for Stability Studies. The data required to perform a transient stability study, and the recommended format for organizing and presenting the information for most convenient use are covered in detail in the application guides for particular stability programs. The following is a summary of the generic classes of data needed. Note, that some of the more esoteric information is not essential; omitting it merely limits the accuracy of the results, especially at times exceeding five times the duration of the disturbance being studied. The more essential items are marked by an asterisk (*).

(1) System data.

(a) Impedances $(R + jX)$ of all significant transmission lines, cables, reactors, and other series components.*

(b) For all significant transformers and autotramsformers

    (i) kVA rating*

    (ii) Impedance*

    (iii) Voltage ratio*

    (iv) Winding connection*

    (v) Available taps and tap in use*

    (vi) For regulators and load tap-changing transformers: regulation range, tap step size, type of tap changer control*

(c) Short circuit capacity (steady-state basis) of utility supply, if any*

(d) kvar of all significant capacitor banks*

(e) Description of normal and alternate switching arrangements*

(2) Load data: real and reactive electrical loads on all significant load buses in the system*

(3) Rotating machine data

(a) For major synchronous machines (or groups of identical machines on a common bus)

(i) Mechanical and/or electrical power ratings (kVA, hp, kW, etc)*

(ii) Inertia constant H or inertia WK^2 of rotating machine and connected load or prime mover*

(iii) Speed*

(iv) Real and reactive loading, if base-loaded generator*

(v) Speed-torque curve or other description of load torque, if motor*

(vi) Direct-axis subtransient,* transient,* and synchronous reactances*

(vii) Quadrature-axis subtransient, transient,* and synchronous reactances

(viii) Direct-axis and quadrature-axis subtransient and transient* time constants

(ix) Saturation information

(x) Potier reactance

(xi) Damping data

(xii) Excitation system type time constants, and limits

(xiii) Governor and steam system or other prime mover type, time constants, and limits

(b) For minor synchronous machines (or groups of machines)

(i) Mechanical and/or electrical power ratings*

(ii) Inertia*

(iii) Speed*

(iv) Direct-axis synchronous reactance*

(c) For major induction machines or groups of machines

(i) Mechanical and/or electrical power ratings*

(ii) Inertia*

(iii) Speed*

(iv) Positive-sequence equivalent circuit data (for example, R_1, X_1, X_M, R_2, X_2)*

(v) Load speed-torque curve*

(vi) Negative-sequence equivalent circuit data

(vii) Description of reduced-voltage or other starting arrangements, if used*

(d) For minor induction machines: detailed dynamic representation not needed, represent as a static load

(4) Disturbance data

(a) General description of disturbance to be studied, including (as applicable) initial switching status; fault type, location and duration; switching operations and timing; manufacturer, type, and setting of protective relays and clearing time of associated breakers*

(b) Limits on acceptable voltage, current, or power swings*

(5) Study parameters

(a) Duration of study*

(b) Integrating interval*

(c) Output printing interval*

(d) Data output required*

7.6.6 Stability Program Output. Most stability programs give the user a wide choice of results to be printed out. The program can calculate and print any of the following information as a function of time:

(1) Rotor angles, torques, and speeds of synchronous machines

(2) Real and reactive power flows throughout the system

(3) Voltages and voltage angles at all buses

(4) System frequency

(5) Torques and slips of all induction machines

The combination of these results selected by the user can be printed out for each printing interval (also user-selected) during the course of the study period.

The value of the study is strongly affected by the selection of the proper printing interval and the total duration of the simulation. Normally a printing interval of 0.01 or 0.02 s is used; longer intervals reduce the computer costs slightly, but increase the risk of missing fast swings of rotor angle. The computer time cost is nearly proportional to the total study time, so this parameter should be closely controlled for the sake of economy.

This is especially important if the system and machines have been represented approximately or incompletely, because the errors will accumulate and render the results meaningless after some point. A time limit of five times the duration of the major disturbance being studied is generally long enough to show whether the system is stable or not, while keeping costs to a reasonable level.

7.6.7 Interpreting Results—Swing Curves. The results of a computer transient stability study are fairly easy to understand once the user learns the basic principles underlying stability problems. The most direct way to determine from the study results whether a system is stable is to look at a set of swing curves for the machines in the system. Swing curves are simply plots of rotor angles against time; if the curves of all the machines involved are plotted on common axes, we can easily see whether they diverge (indicating instability) or settle to new steady-state values.

For example, Fig 62 shows swing curves for a system which is stable under the disturbance applied. This is a reproduction of an actual computer printout. A simplified one-line diagram of the system appears as Fig 63. Note that while the three-phase bolted fault on a synchronous bus feeder, Case I, cleared by instantaneous tripping of the feeder breaker, causes all five generators to experience swings of varying magnitude, the oscillations in the rotor angles are obviously damped and can be expected to die out.

By contrast, in Case II the fault is applied to the tie between the synchronizing bus and one of the generator buses and is cleared by tripping the tie circuit breaker. The swing curves for this condition are shown in Fig 64. Generator No 1 is disconnected from the system and suffers a severe overload, causing it to decelerate, as shown by a unidirectional negative change in rotor angle. The other machines stay in synchronism.

7.7 Stability Studies on a Typical System. Probably the best way to examine some of the typical applications of stability analysis to industrial power systems is to look at the stability studies which would go into the design of a typical large industrial system including 20 MVA of local generation and 40 MVA of purchased power capacity. The stability studies which might be applied to this system are:

(1) The basic layout of the primary system can be affected by stability considerations. For example, an initial design choice might be to connect the generated and purchased power buses through only one tie circuit. However, stability studies could show that inadequate synchronizing power is available to prevent the generators and the utility from losing synchronism during primary system faults unless two ties are provided. The same sort of considera-

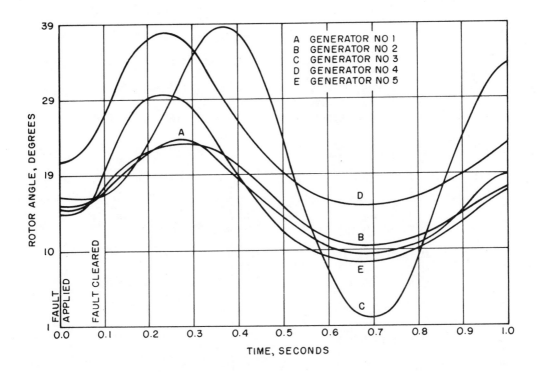

Fig 62
Computer Printout of Swing Curves for
Case I Fault on System in Fig 63

tions might dictate that the 4160 V bus ties be operated closed, to ensure the lowest possible impedance between the synchronous motors and the power sources to enhance stability.

(2) Related to the design of the basic layout is the problem of protective relaying. The system can be designed for maximum inherent stability by closely coupling all machines. Or the same objective can often be obtained by designing the protective scheme for the fastest possible clearing of faults. Since the former choice may involve economic sacrifices in the form of higher capacity switchgear, often the latter choice represents the best solution. Extra-fast relaying can conflict with the requirements of a selectively coordinated system, however, unless expensive zone protection schemes (bus differential, pilot wire, etc) are used. Balancing all of these factors, probably the best procedure is to design the system layout around process requirements, provide the fastest relaying possible within the constraints of selectivity and economics, and then check the proposed layout and relaying by a series of stability studies simulating the more probable fault conditions. In the system shown in Fig 65, three-phase faults are applied on one 138 kV utility line ahead of the plant transformers, on a feeder from each of the 13.8 kV buses, and on a feeder from each 4160 V bus. Of course, the simulation would include

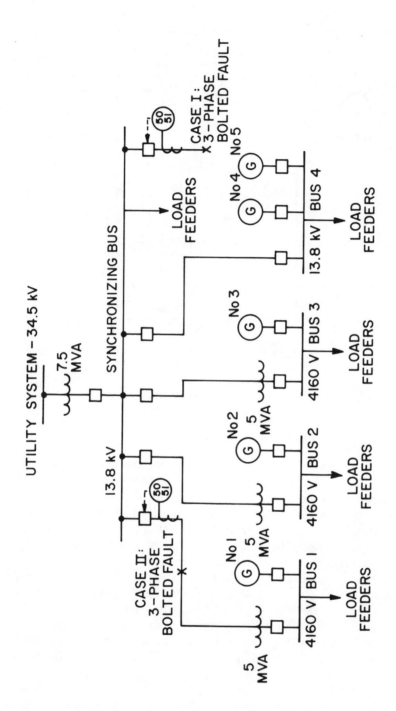

Fig 63
Single-Line Diagram of System whose
Swing Curves Appear in Figs 62 and 64

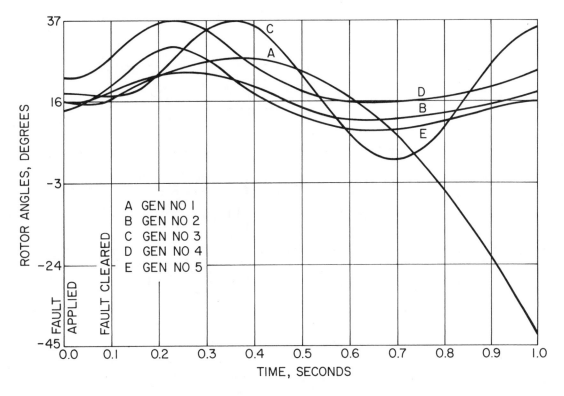

Fig 64
Computer Printout of Swing Curves for
Case II Fault on System Shown in Fig 63

clearing of the fault via the proposed relaying. If any of these studies show an unstable condition, further stability studies might be required to test the effectiveness of various proposed solutions.

(3) In the system shown in Fig 65 some considerations should certainly be given to automatic load shedding. If the power company suffers an outage on the 138 kV lines while the plant is running at nearly full load, the 20 MVA of local generation will abruptly be subjected to an overload approaching 300%. This overload will promptly cause the generators to trip off, leaving the plant with no power at all, even though 20 MVA of perfectly sound generation is available to maintain service to the most critical loads. Obviously, a method of automatically interrupting noncritical loads commensurately with the loss of system capacity would be valuable.

One such possibility would be to trip noncritical feeders whenever the utility tie breakers are opened. However, this wired-in scheme lacks flexibility. To permit shedding only the amount of load required to prevent system collapse, many industrial plants with local generation use underfrequency relaying. This scheme depends on the fact that an overloaded generator slows down, dropping the system frequency.

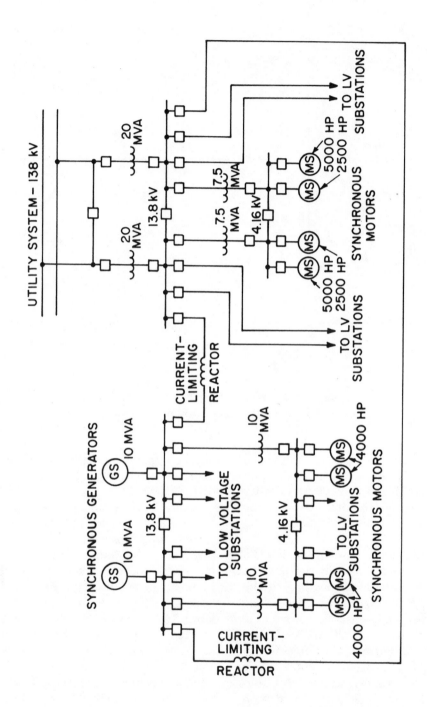

Fig 65
Single-Line Diagram of a Typical Large Industrial
Power System with On-Site Generation

A two-stage load shedding scheme might operate as follows: The first-stage relay operates at 59 Hz and a time delay of 6 cycles, tripping 10 or 15 MVA of noncritical load; the second stage operates at 58 Hz and a delay of 30 cycles, tripping an additional 20 MVA of somewhat more critical load.

In designing the load-shedding scheme for the system in Fig 65, first run a stability study to calculate the decay of system frequency when the utility tie is lost, without any load shedding. Using this frequency decrement curve, estimates can be made of the amounts of load to be shed and the frequency and time delay settings for the underfrequency relays. Then these data can be used in the stability study program to calculate the system frequency versus time curve with the proposed load shedding. If sufficient load is shed fast enough to prevent system collapse, the validity of the proposed relay scheme and settings is confirmed. Usually several runs are made with different system conditions in each load shedding analysis.

(4) In the system shown in Fig 65, the effect of starting one of the large synchronous motors could be substantial, especially under abnormal conditions when one or more power sources are out of service. This effect can be evaluated by a series of stability studies simulating motor starting under various conditions of system capacity, prestart motor terminal voltage, etc. The study results yield motor accelerating times, real and reactive power flows, and bus voltages at all critical points in the system.

As this discussion indicates, transient stability analysis should be an integral part of the design or expansion of any industrial power system containing several large synchronous machines, and should be considered even if only one machine is being applied.

7.8 References

[1] *Electrical Transmission and Distribution Reference Book*, Westinghouse Electric Corporation, East Pittsburgh, PA., 1964, chap. 13.

[2] KIMBARK, E. W. Power System Stability, vol 1, New York, John Wiley, 1948.

[3] FITZGERALD, A. E. and KINGSLEY, CHARLES Jr. Electric Machinery, New York, McGraw-Hill, 1961, chap 5.

8. Motor Starting Studies

8.1 Introduction. This section discusses benefits obtained from motor starting studies and examines various types of computer-aided studies normally performed. Data or information required to permit these studies along with expected results of a motor starting study effort are also reviewed.

8.2 Need for Motor Starting Studies

8.2.1 Problems Revealed. Motors on modern industrial systems are becoming increasingly larger. Some are considered large even in comparison to the total capacity of large industrial power systems. Starting large motors, especially across-the-line, can cause severe disturbances to the motor and any locally connected load, and also to buses electrically remote from the point of motor starting. A brief discussion of major problems associated with starting large motors, and therefore of significance in power system design and evaluation, follows.

8.2.2 Voltage Dips. Probably the most widely recognized and studied effect of motor starting is the voltage dip experienced throughout an industrial power system as a direct result of starting large motors. Available accelerating torque drops appreciably at the motor bus as voltage dips to a lower value, extending the starting interval and affecting, sometimes adversely, overall motor starting performance. During motor starting, voltage level at the motor terminals should be maintained at approximately 80% of rated voltage or above for a standard National Electrical Manufacturers Association (NEMA) type B motor. This value results from examination of speed-torque characteristics of this type motor (150% starting torque at full voltage) and the desire to successfully accelerate a fully loaded motor at reduced voltage (that is, torque varies with the square of the voltage). When other motors or lower shaft loadings are involved, the speed-torque characteristics of both the motor and its load should be examined to specifically determine minimum acceptable voltage. Assuming reduced voltage permits adequate accelerating torque, it should also be verified that the longer starting interval required at reduced torque caused by a voltage dip does not result in the I^2t damage limit of the motor being exceeded.

8.2.3 Weak Source Generation. Smaller power systems are usually served by limited capacity sources, which generally magnify voltage drop problems on motor starting, especially when large motors are involved. Small systems also often have limited on-site generation, which further complicates normal problems since additional voltage drops occur in transient impedances of local generators during the motor starting interval. The type of voltage regulator system applied with the generators can dramatically influence motor starting as illustrated in Fig 66. A motor starting study can be useful, even for analyzing the performance of small systems. Certain digital computer programs can accurately model exciter/regulator response under motor starting conditions necessary for meaningful results and conclusions.

8.2.4 Special Torque Requirements. Sometimes special loads must be accelerated under carefully controlled conditions without exceeding specified torque limitations of the equipment. An example of this is starting a motor connected to a load through gearing. This application requires a special period of low torque *cushioned* acceleration to allow slack in the gears and couplings to be picked up without damage to the equipment. Certain computer-aided motor starting studies allow an instant-by-instant shaft output torque tabulation for comparison to allowable torque limits of the equipment. This study can be used for selecting a motor or a starting method, or both, with optimum speed-torque characteristics for the application. The results of a detailed study are used for sizing the starting resistors for a wound rotor motor or in analyzing rheostatic control for a starting wound rotor motor which might be used in a *cushioned* starting application involving mechanical gearing or a coupling system that has torque transmitting limitations. High inertia loads increase motor starting time, and heating in the motor due to high currents drawn

Fig 66
Typical Generator Terminal Voltage Characteristics
for Various Exciter Regulator Systems

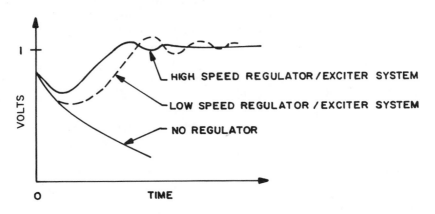

during starting can be intolerable. A computer-aided motor starting study allows accurate values of motor accelerating current and time. This makes it possible to determine if thermal limits of standard motors will be exceeded for longer than normal starting intervals.

Other loads have special starting torque requirements or accelerating time limits that require special *high* starting torque (and inrush) motors. Additionally, the starting torque of the load or process may not permit low inrush motors in situations where these motors might reduce the voltage dip caused by starting a motor having standard inrush characteristics. A simple inspection of the motor and load speed-torque curves is not sufficient to determine whether such problems exist. This is another area where the motor torque and accelerating time study can be useful.

8.3 Recommendations

8.3.1 Voltage Dips. A motor starting study can expose and identify the extent of a voltage drop problem. The voltage at each bus in the system can, for example, be readily determined by a digital computer study. Equipment locations likely to experience difficulty during motor starting can be immediately determined.

In situations where a variety of equipment voltage ratings are available, the correct rating for the application can be selected. Circuit changes such as off-nominal tap settings for distribution transformers and larger than standard conductor size cable can also be readily evaluated. On a complex power system, this type of detailed analysis is very difficult to accomplish with time-consuming hand solution methods.

Several methods of minimizing voltage dip on starting motors are based on the fact that during starting time, a motor draws an inrush current directly proportional to terminal voltage, and therefore a lower voltage causes the motor to require less current, thereby reducing the voltage dip. Auto-transformer starters are a very effective means of obtaining a reduced voltage during starting with standard taps ranging from 50% to 80% of normal rated voltage. A motor starting study is used to select the proper voltage tap and the lower line current inrush for the electrical power system during motor start. Other special reduced voltage starting methods include resistor or reactor starting, part-winding starting, and wye (Y)-start delta (Δ)-run motors. All are examined by an appropriate motor starting study and the best method for the particular application involved can be selected. In all reduced voltage starting methods, torque available for accelerating the load is a very critical consideration once bus voltage levels are judged otherwise acceptable. Only 25% torque is available, for example, with 50% of rated voltage applied at the motor terminals. Any problems associated with reduced starting torque imposed by special starting methods are automatically uncovered by a motor starting study.

Another method of reducing high inrush currents when starting large motors is a capacitor starting system [1]. This maintains acceptable voltage levels throughout the system. With this method, the high inductive component of normal reactive starting current is offset by the addition, during the starting period only, of capacitors to the motor bus. This differs from the practice of applying capacitors for running motor power factor correction. A motor starting study can provide information to allow optimum sizing of the starting capacitors and determination of the length of time the capacitor must be energized. The study can also establish

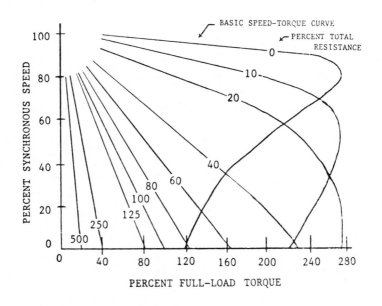

Fig 67
Typical Wound Rotor Motor Speed-Torque Characteristics

whether the capacitor and motor can be switched together, or because of an excessive voltage drop that might result from the impact of capacitor transient charging current when added to the motor inrush current the capacitor must be energized momentarily ahead of the motor. The switching procedure can appreciably affect the cost of final installation.

Use of special starters or capacitors to minimize voltage dips can be an expensive method of maintaining voltage at acceptable levels [1]. Where possible, off-nominal tap settings for distribution transformers are an effective, economical solution for voltage dips. By raising no-load voltage in areas of the system experiencing difficulties during motor starting, voltage dip can often be minimized. In combination with a load flow study, a motor starting study can provide information to assist in selecting proper taps and ensure that light-load voltages are not excessively high.

The motor starting study can be used to prove effectiveness of several other solutions to the voltage dip problem as well. With a wound rotor motor, differing values of resistance are inserted into the motor circuit at various times during the starting interval to reduce maximum inrush (and accordingly starting torque) to some desired value. Figure 67 shows typical speed-torque characteristic curves for a wound rotor motor. With appropriate switching times (dependent on motor speed) of resistance values, practically any desired speed-torque (starting) characteristic can be obtained. A motor starting study aids in choosing optimum current and torque values for a wound rotor motor application whether resistances are switched in steps by timing relays or continuously adjusted values obtained through a liquid rheostat-feedback starting control.

8.3.2 Analyzing Starting Requirements. A speed-torque and accelerating time study often in conjunction with the previously discussed voltage dip study permits a means of exploring a variety of possible motor speed-torque characteristics. This type of motor starting study also confirms that starting times are within acceptable limits. The accelerating time study assists in establishing the necessary thermal damage characteristics of motors or verifies that machines with locked rotor protection supervised by speed switches will not experience nuisance tripping on starting.

Speed-torque/accelerating time motor starting study is also used to verify special motor torque or inrush characteristics, or both, specified to actually produce desired results. Mechanical equipment requirements and special ratings necessary for motor starting auxiliary equipment are based on information developed from a motor starting study.

8.4 Types of Studies. From the above discussion, it is apparent that depending on the factors of concern in any specific motor starting situation, more than one type of motor starting study can be required.

8.4.1 The Voltage Drop Snapshot. One method of examining the effect of voltage dip during motor starting is to ensure the maximum instantaneous drop that occurs leaves bus voltages at acceptable levels throughout the system. This is done by examining the power system that corresponds to the worst case voltage. Through appropriate system modeling, this study can be performed by various calculating methods using the digital computer.

The *snapshot* voltage drop study is useful only for finding system voltages. Ex-

cept for the recognition of generator transient impedances when appropriate, machine inertias, load characteristics, and other transient effects are usually ignored. This type of study, while certainly an approximation, is often sufficient for many applications.

8.4.2 The Detailed Voltage Profile. This type of study allows a more exact examination of the voltage drop situation. Regulator response, exciter operation, and sometimes governor action are modeled to accurately represent transient behavior of local generators. This type of study is similar to a simplified transient stability analysis and can be considered a series of voltage *snapshots* throughout the motor starting interval including the moment of minimum or *worst* case voltage.

8.4.3 The Motor Torque and Acceleration Time Analysis. Perhaps the most exacting analysis for motor starting conditions is the detailed speed-torque analysis. Similar to a transient stability study (some can also be used to accurately investigate motor starting), speed-torque analysis provides electrical and accelerating torque calculations for specified time intervals during the motor starting period. Motor slip, load and motor torques, terminal voltage magnitude and angle, and the complex value of motor current drawn are values to be examined at time zero and at the end of each time interval.

Under certain circumstances, even across-the-line starting, the motor may not be able to *break away* from standstill or it may stall at some speed before acceleration is complete. A speed-torque analysis, especially when performed using a computer program, and possibly in combination with one or more previously discussed studies, can predict these problem areas and allow corrections to be made before difficulties arise. When special starting techniques are necessary, such as auto-transformer reduced voltage starting, speed-torque analysis can account for the auto-transformer magnetizing current and it can determine the optimum time to switch the transformer out of the circuit. The starting performance of wound rotor motors is examined through this type of study.

8.4.4 Adaptations. A particular application can require a slight modification of any of the above studies to be of greatest usefulness. Often combinations of several types of studies described are required to adequately evaluate system motor starting problems.

8.5 Data Requirements

8.5.1 Basic Information. Since other loads on the system during motor starting affect voltage available at the motor terminals, essentially the same information necessary for a load flow or short circuit study is also required for a motor starting study. Although this information is summarized below, details are available elsewhere in this standard (see Sections 6 and 7) as well as in [2] and [3].

(1) *Utility and Generator Impedances.* These values are extremely significant and should be as accurate as possible. Generally they are obtained from local utility representatives and generator manufacturers. Where exact generator data cannot be obtained, typical impedance values are available from [2] and [3].

(2) *Transformers.* Manufacturers' impedance information should be obtained where possible, especially for large units (that is, 5000 kVA and larger). Standard impedances can usually be used with little error for smaller units, and typical X/R ratios are available in ANSI/IEEE

C37.010-1979 [4].

(3) *Other Components.* System elements (such as cables) should be specified as to the number and size of conductor, conductor material, and whether magnetic duct or armor is used. All system elements should be supplied with R and X values so an equivalent system impedance can be calculated.

(4) *Load Characteristics.* System loads should be detailed including type (constant current, constant impedance, or constant kVA), power factor, and load factor, if any. Exact inrush (starting) characteristics should also be given for the motor to be started.

(5) *Machine and Load Data.* Aside from base information required for a voltage drop type of motor starting analysis, several other items are required for the detailed speed-torque and accelerating time analysis. These include the WR^2 of the motor and load (with the WR^2 of the mechanical coupling or any gearing included), and speed-torque characteristics of both the motor and load. Typical speed-torque curves are shown in Fig 68.

**Fig 68
Typical Motor and
Load Speed-Torque
Characteristics**

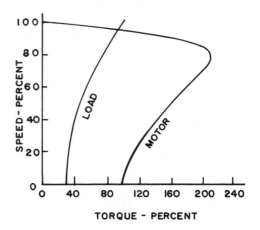

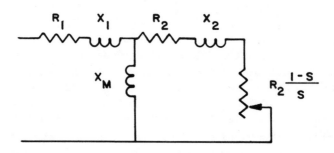

Fig 69
Simplified Equivalent Circuit for a
Motor on Starting

For additional accuracy, speed versus current, and speed versus power factor characteristics should be given for as exact a model as possible for the motor during starting. For some programs, constants for the motor equivalent circuit given in Fig 69 can be either required or alternatively utilized as input information. This data must be obtained from the manufacturer since values are critical.

8.5.2 Simplifying Assumptions. Besides using standard impedance values for transformers and cables, it is often necessary to use typical or assumed values for other variables when making motor starting voltage drop calculations. This is particularly true when calculations are for evaluating a preliminary design and exact motor and load characteristics are unknown. Some common assumptions used in the absence of more precise data follow:

(1) *Horsepower to kVA Conversion.* A reasonable assumption is 1 hp equals 1 kVA. For induction motors and synchronous motors with 0.8 leading, running power factor, it can easily be seen from the equation:

$$\text{hp} = \frac{(\text{kVA})\,(0.746)}{(\text{EFF})\,(\text{PF})} \quad \text{(See Ref [5].)}$$

The ratio of 0.746 to efficiency times the power factor approaches unity for most motors giving the 1 hp/kVA approximation. Therefore, for synchronous motors operating at 1.0 PF, a reasonable assumption is 1 hp equals 0.8 kVA.

(2) *Inrush Current.* Usually, a conservative multiplier for motor starting inrush currents is obtained by assuming the motor to have a code G characteristic with locked rotor current equal to approximately 6 times the full load current with full voltage applied at motor terminals, [6].

(3) *Starting Power Factor.* The power factor of a motor during starting determines the amount of reactive current that is drawn from the system, and thus to a large extent the maximum voltage drop. Typical data [2] suggest the following:

(a) Motors under 1000 hp, PF = 0.20
(b) Motors 1000 hp and over, PF = 0.15

8.6 Solution Procedures and Examples. Regardless of the type of study required, a basic voltage drop calculation is always involved. When voltage drop is the only concern, the end product is this calculation when all system impedances are at maximum value and all voltage sources are at minimum expected level. In a more complex motor speed-torque analysis and accelerating time study, voltage drop calculations are required. These are performed at regular time intervals following the initial impact of the motor starting event and take into account variations in system impedances and voltage sources. Results of each *iterative* voltage drop calculation are used to calculate output torque which is dependent on the voltage at machine terminals and motor speed. Since the interval of motor starting usually ranges from a few seconds to 10 or more seconds, effects of generator voltage regulator and governor action are evident, sometimes along with transformer tap control depending on control settings. Certain types of motor starting studies account for generator voltage regulator action while a transient stability study is usually required in cases where other transient effects are considered important. A summary of fundamental equations used in various types of motor starting studies follows, along with examples illustrating applications of fundamental equations to actual problems which illustrate typical computer program outputs.

8.6.1 The Mathematical Relationships. There are basically three ways to solve for bus voltages realized throughout the system on motor starting. These are presented in this section and then examined in detail by examples in the next section.

(1) *Impedance Method.* This method involves reduction of the system to a simple voltage divider network [7] where voltage at any point (bus) in a circuit is found by taking known voltage (source bus) times the ratio of impedance to the point in question over total circuit impedance. For the circuit of Fig 70,

$$V = E \frac{X_1}{X_1 + X_2}$$

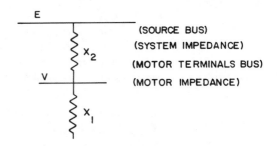

(SOURCE BUS)
(SYSTEM IMPEDANCE)
(MOTOR TERMINALS BUS)
(MOTOR IMPEDANCE)

**Fig 70
Simplified Impedance
Diagram**

or, more generally,

$$V = E \frac{Z_1}{Z_1 + Z_2}$$

The effect of adding a large capacitor bank at the motor bus is seen by the above expression for V. The addition of negative vars causes X_1, or Z_1, to become larger in both numerator and denominator so bus voltage V is increased and approaches 1.0 per unit as the limiting improvement. Locked rotor impedance for three-phase motor is simply

$$Z_{LR} = \frac{\text{rated voltage L-L}}{(\sqrt{3} \ (I_{LR})} \text{ in } \Omega$$

where

I_{LR} = locked rotor current at rated voltage

This value in per unit is equal to the inverse of the inrush multiplier on the motor rated kVA base:

$$Z_{LR} = \frac{1}{I_{LR}/I_{FL}} \text{ in per unit}$$

Since a starting motor is accurately represented as a constant impedance, the im-

147

pedance method is a very convenient and acceptable means of calculating system bus voltages during motor starting. Validity of the impedance method can be seen and is usually used for working longhand calculations. Where other than simple radial systems are involved, the digital computer greatly aids in obtaining necessary network reduction. To obtain results with reasonable accuracy, however, various system impedance elements must be represented as complex quantities rather than as simple reactances.

(2) *Current Method.* For any bus in the system represented in Figs 71 and 72, the basic equations for the current method are as follows:

$$I_{\text{per unit}} = \frac{\text{MVA}_{\text{load}}}{\text{MVA}_{\text{base}}} \text{ at } 1.0 \text{ per unit voltage}$$

$$V_{\text{drop}} = I_{\text{per unit}} \cdot Z_{\text{per unit}}$$

$$V_{\text{bus}} = V_{\text{source}} - V_{\text{drop}}$$

The quantities involved should be expressed in complex form for greatest accuracy although reasonable results can be obtained by using magnitudes only for first order approximations.

The disadvantage to this method is that since all loads are not of constant current type, the current to each load varies as voltage changes. An iterative type solution procedure is therefore necessary to solve for the ultimate voltage at every bus, and such tedious computations are readily handled by a digital computer.

(3) *Load Flow Solution Method.* From the way loads and other system elements are portrayed in Figs 71 and 72, it appears that bus voltages and the voltage dip could be determined by a conventional load flow program. This is true. By modeling the starting motor as a constant impedance load, the load flow calculations yield the bus voltages during starting. The basic equations involved in this process are repeated here [8], [9].

**Fig 71
Typical One-Line Diagram**

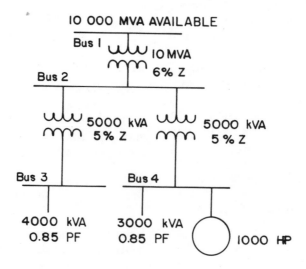

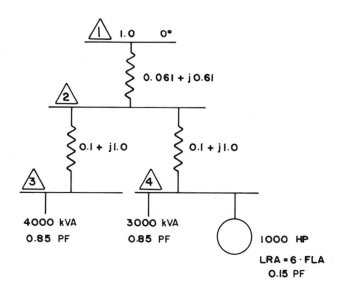

LRA = LOCKED ROTOR CURRENT

FLA = FULL LOAD CURRENT

Fig 72
Impedance Diagram for System
in Fig 71

$$I_k = \frac{P_k - jQ_k}{V_k{}^*} - Y_k V_k$$

$$V_k = V_{\text{ref}} + \sum_{i = 1}^{n} Z_{ki}\left(\frac{P_i - jQ_i}{V_i} - Y_i V_i\right)$$

where

I_k = current in the kth branch of
any network

P_k, Q_k = real and reactive powers representative of the loads at the kth bus

V_k = the voltage at the kth bus

Y_k = the admittance to ground of
bus k

V_{ref} = voltage of the swing or slack bus

n = number of buses in the network

Z_{ki} = the system impedance between the kth and ith buses

The load flow solution to the motor starting problem is very precise for finding bus voltages at the instant of maximum voltage drop. It is apparent from the expressions for I_k and V_k that this solution method is ideally suited for the digital computer any time the system involves more than two or three buses.

8.6.2 Other Factors. Unless steady-state conditions exist, all of the above solution methods are valid for one particular instant and provide the single *snapshot*

149

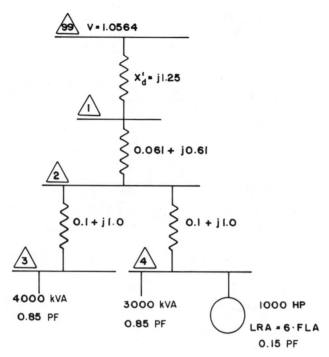

Fig 73
Revised Impedance Diagram Showing
Transient Reactance of Generator

of system bus voltages as mentioned earlier. For steady-state conditions it is assumed that generator voltage regulators have had time to increase field excitation sufficiently to maintain the desired generator terminal voltage. Accordingly, the presence of the internal impedance of any local generators connected to the system is ignored. During motor starting, however, the influence of machine transient behavior becomes important. To model the effect of a close-connected generator on the maximum voltage drop during motor starting requires inclusion of generator transient reactance in series with other source reactances. In general,

use of the transient reactance as the representation for the machine results in calculated bus voltages and, accordingly, voltage drops, that are reasonably accurate and conservative, even for exceptionally slow-speed regulator systems.

Assuming for example that bus No 7 in the system shown in Fig 72 is at the line terminals of a 12 MVA generator rather than being an infinite source ahead of a constant impedance utility system, the transient impedance of the generator would be added to the system. The resulting impedance diagram is shown in Fig 73. A new bus 99 is created. Voltage at this new bus is frequently referred to

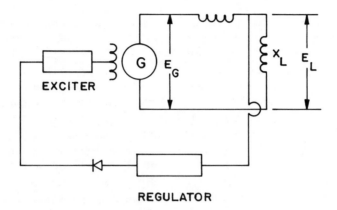

$$E_L = E_G \frac{(X_L)}{X_L + X_G}$$

where

X_G varies with time as
$X_d'' \to X_d' \to X_d$, and E_G varies with the time
constants $T_{do}'' \to T_{do}' \to T_{do}$
depending on exciter/regulator output

**Fig 74
Simplified Representation of Generator
Exciter/Regulator System**

as *voltage-behind-the-transient-reactance*. It is actually the internal machine transi-end driving voltage (see Section 3).

When steady-state operating voltage is 1.0 per unit, it is considered the voltage that must be present ahead of the generator transient reactance prior to supplying power to the other loads on the system and maintain terminal voltage at 1.0 per unit (within exciter tolerances) during steady-state conditions. The transient driving voltage V is calculated as follows:

$$V = V_{terminal} + (jX_d)(I_{load})$$
$$= 1 + (jX_d)(I_{load})$$

when

$$V_T = 1.0 \text{ per unit}$$

and, where

$$I_{load} = \frac{MVA_{load}}{MVA_{base}} \text{ per unit}$$

Treatment of a locally connected generator is equally applicable to all three solution methods described above. Such an approach cannot give any detail regarding the response of the generator voltage regulator or changes in machine characteristics with time. For a more detailed solution which considers time dependent effects of machine impedance and voltage regulator action, the appropriate impedance and voltage terms in each expression must be continuously altered to accurately reflect changes which occur in the circuit. This procedure is also applicable to any solution methods considered. Figure 74 shows a simplified representation of the machine parameters which must be repeatedly modified to obtain the correct solution.

Some type of reduced voltage starting is often used to minimize motor inrush current and thus reduce total voltage drop, when the associated reduction in torque accompanying this starting method is permissible. Representation used for the motor in any solution method for calculating voltage drop must be modified to reflect the lower inrush current. If auto-transformer reduced voltage starting is used, motor inrush will be reduced by the appropriate factor from Table 12. If, for example, normal inrush is six times full load current and an 80% tap auto-transformer starter is applied,

the actual inrush multiplier used for determining the appropriate motor representation in the calculations is (6) (0.64) = 4.2 times full load current.

Resistor or reactor starting limits the line starting current by the same current as motor terminal voltage is reduced (that is, 65% of applied bus voltage gives 65% of normal line starting current).

Y-start, Δ-run starting delivers 33% of normal starting line current with full voltage at the motor terminals. The starting current at any other voltage is, correspondingly, reduced by the same amount. Part winding starting allows 60% of normal starting line current at full voltage and reduces inrush accordingly at other voltages.

When a detailed motor speed-torque and accelerating time analysis is required, the following equations found in many texts apply [10]. The equations in general apply to both induction and synchronous motors since the latter behave almost exactly as do induction machines during the starting period.

$$T \propto V^2$$

$$T = I_0 \alpha$$

$$I_0 = \frac{WK^2}{g} \text{ lb-ft-s}^2$$

Table 12
Auto-Transformer-Line
Starting Current

Autotransformer Tap (% Line Voltage)	Line Starting Current (% Normal at Full Voltage)
50	25
65	42
80	64

$$\omega^2 = \omega_0{}^2 + 2\alpha \, (\theta - \theta_0) \text{ r/s}$$

$$\Delta\theta = \omega_0 t + \tfrac{1}{2}\,\alpha\, t^2 \text{ r}$$

$$\alpha = \frac{T_n \, 2g}{WR^2} \text{ r/s}^2$$

A simplified approximation for starting time is also available:

$$t \text{ (s)} = \frac{WK^2 \, (\text{r/min}_1 - \text{r/min}_2) \, (2\pi)}{60 g T_n}$$

where

T = average motor shaft output torque

V = motor terminal voltage

I_0 = moment of inertia

g = acceleration due to gravity

ω = angular velocity

α = angular acceleration

t = time in seconds to accelerate

T_n = net average accelerating torque between rpm$_1$ and rpm$_2$

θ = electrical angle in degrees

WK^2 = machine inertia constant

The basic equation for use with the equivalent circuit of Fig 69 is as follows [2], [11], [12]:

$$T = \frac{1}{\omega_s} \frac{q_1 V^2 \, (r_2/s)}{(r_1 + r_2/s)^2 + (X_1 + X_2)^2}$$

where

T = instantaneous torque

ω_s = angular velocity at synchronous speed

$(r_1 + jX_1)$ = stator equivalent impedance

$(r_2/s + jX_2)$ = rotor equivalent impedance

q_1 = number of stator phases (3 for a 3 ϕ machine)

V = motor terminal voltage

8.6.3 The Simple Voltage Drop Determination.

To illustrate this type of a computer analysis, the system of Fig 71 will again be considered. It is assumed that bus No 1 is connected to the terminals of a 12 MVA generator having 15% transient reactance (1.25 per unit on a 100 MVA base). Prior to starting, when steady-state load conditions exist, the impedance diagram of Fig 72 applies with the motor disconnected. The impedance diagram of Fig 73 applies when the 1000 hp motor on bus 4 is started.

Bus 99 in Fig 73 has been assigned in voltage of 1.056 per unit. This value can be confirmed using the expression for the internal machine transient driving voltage V given in the previous section with appropriate substitutions as follows:

$$V = (1.0 + j0.0) + (j1.25) \, (0.060114 + j0.042985)$$
$$= 1.0564 \text{ per unit voltage } \angle \, 4.08°$$

where values for the current through the X_d' element correspond to those that exist at steady state prior to motor starting with generator bus 1 operating at 1.0 per unit. The computer output report shown in Fig 75 shows steady-state load flow results for this case. All system loads are connected except the 1000 hp motor. Power flows are expressed in MW and Mvar. Quantities indicated for real and reactive power flow (and, accordingly, the circuit flow at 1.0 per unit voltage) through the line between bus 1 and bus 2 agree exactly with those used above to calculate V. Likewise, the bus 1 voltage is shown to be 1.0 per unit while the bus 99 voltage is 1.0564 per unit.

For convenience, the angle associated with the bus 99 reference voltage is assumed to be zero, which simply results in a corresponding shift in all other bus

NO.	BUS NAME	LINE FROM - TO	VOLTAGE (P.U.)	ANGLE (DEGREES)	NET MEGAWATTS	NET MEGAVARS	NET MVA	TAP RATIO (P.U.)	CKT NO.	INTERMEDIATE MISMATCH
1	SWING		1.0000	0.0	6.0114	4.2985				
		1 - 2			6.0114	4.2985	7.3901	0.0	1	0.0 0.0
2	MAIN XFMR SEC		0.9707	-2.01	0.0	0.0				
		2 - 1			-5.9781	-3.9654	7.1736	0.0	1	
		2 - 3			3.4180	2.2865	4.1122	0.0	1	-0.000 0.001
		2 - 4			2.5599	1.6795	3.0617	0.0	1	
3	SMVA XFMR SEC		0.9442	-4.00	-3.4000	-2.1070				
		3 - 2			-3.4000	-2.1070	4.0000	0.0	1	-0.000 -0.000
4	MTR STRT BUS		0.9511	-3.49	-2.5500	-1.5800				
		4 - 2			-2.5500	-1.5800	2.9998	0.0	1	0.000 -0.000

LINES	BUSES	LTCS	TOTAL-MISMATCH			
3	4	0	-0.000 0.001			
			GENERATION	6.0114	4.2985	
			LOAD	5.9500	3.6870	
			SYSTEM LOSSES	0.0612	0.6121	ITERATIONS 9

Fig 75
Load Flow Computer Output (steady state)

NO.	BUS NAME	LINE FROM - TO	VOLTAGE (P.U.)	ANGLE (DEGREES)	NET MEGAWATTS	NET MEGAVARS	NET MVA	TAP RATIO (P.U.)	CKT NO.	INTERMEDIATE MISMATCH	
1	MAIN XFMR PRI		0.9292	-4.89	0.0	0.0					
		1 - 99			-6.6904	-9.1669	11.3487	0.0	1	0.001	0.001
		1 - 2			6.6911	9.1681	11.3501	0.0	1		
2	MAIN XFMR SEC		0.8655	-7.40	0.0	0.0					
		2 - 1			-6.6001	-8.7381	10.5715	0.0	1	0.001	0.001
		2 - 3			3.4230	2.3364	4.1444	0.0	1		
		2 - 4			3.1778	5.9227	6.7214	0.0	1		
3	5MVA XFMR SEC		0.8354	-9.93	-3.4000	-2.1070					
		3 - 2			-3.4001	-2.1071	4.0001	0.0	1	-0.000	-0.000
4	MTR STRT BUS		0.7940	-9.55	-3.1173	-5.1195					
		4 - 2			-3.1175	-5.3196	6.1658	0.0	1	-0.000	-0.000
99	SWING		1.0564	0.0	6.6904	11.0313					
		99 - 1			6.6904	11.0313	12.9015	0.0	1	0.0	0.0

LINES 4	BUSES 5	LTCS 0	TOTAL-MISMATCH 0.001 0.002	GENERATION 6.6904 11.0313	LOAD 6.5173 7.4265	SYSTEM LOSSES 0.1742 3.6068	ITERATIONS 30

Fig 76
Load Flow Computer Output (voltage dip on motor starting)

155

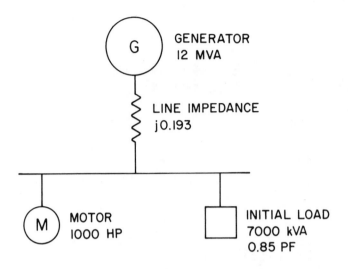

Fig 77
Simplified System Model for
Generator Representation During Motor Starting

voltages. It is seen from the motor starting bus voltage computer report in Fig 76 that when this representation is used and subsequent motor starting calculations are made, the voltage at bus No 4 is 0.7940 $\angle$ –9.55° per unit. This voltage is well below the 0.85 criterion established earlier for proper operation of ac control devices.

8.6.4 Time-Dependent Bus Voltages. The load flow solution method for examining effects of motor starting allows a look at the voltage on the various system buses at a single point in time. A more exact approach is to model generator transient impedance characteristics and voltage sources closer to give results for a number of points in time following the motor starting event. Although the solution methods are applicable to multiple generator/motor systems as well, equations can be developed for a system of the form shown in Fig 77 to solve for generator, motor, and exciter field voltages as a function of

time. The digital computer is used to solve several simultaneous equations that describe the voltage of each bus in a system at time zero and the end of successive time intervals.

Figures 78–81 show in detail the type of input information required and the output obtained from a digital computer voltage drop study. The system shown in Fig 77 contains certain assumptions which include the following:

(1) Circuit losses are negligible—reactances only used in calculations

(2) Initial load is constant kVA type

(3) Motor starting load is constant impedance type

(4) Motor starting power factor is in the range 0 to 0.25

(5) Mechanical effects such as governor response, prime mover speed changes, and inertia constants are negligible

Plotted results obtained from the computer compare favorably to those expected from an examination of Fig 66. In the particular computer program used

PROGRAM ASSUMPTIONS ----
1. ONLY REACTANCES ARE USED IN CALCULATIONS.
2. EXCITER RESPONSE IS MODELED LINEARLY.
3. GENERATOR SATURATION IS MODELED WITH FROHLICH'S EQUATION.
4. THE INITIAL LOAD IS A CONSTANT KVA TYPE.
5. MOTOR STARTING LOAD IS A CONSTANT IMPEDANCE TYPE.
6. MOTOR STARTING P.F. IS IN THE RANGE 0.00 TO 0.25.
7. MECHANICAL EFFECTS (LIKE GOVERNOR RESPONSE, PRIME MOVER SPEED CHANGES, ETC.) ARE NEGLECTED.

PROGRAM DATA BASE ----
GENERATOR KVA= 12000.00
XD=1.300
X'D=0.150
XL=0.120
T'DO= 5.000
IFAG= 0.395
IFNL= 0.450
IF130= 0.692
IFFL= 1.000
C=0.750

STATIC TYPE EXCITER
CEILING VOLTAGE=3.60 P.U.
SET POINT=1.00 P.U.
EXCITER TIME CONSTANT=0.32
REGULATOR TIME CONSTANT=0.0

REACTANCE BETWEEN GENERATOR AND BUS= 0.193

HORSEPOWER OF STARTING MOTOR= 1000.0
INRUSH CURRENT IN P.U. OF RATED=6.00
RATED VOLTAGE= 1.
OPERATING VOLTAGE= 1.

INITIAL LOAD KVA= 7000.00
P.F.=0.85

NOTE ----
ANSWERS GIVEN BELOW REFLECT THE TIME REQUIRED FOR THE GENERATOR VOLTAGE TO RECOVER FROM THE IMPACT OF THE MOTOR STARTING LOAD. THEY DO NOT CORRESPOND TO THE TIME REQUIRED FOR THE MOTOR TO START, NOR DO THEY IMPLY THAT SUFFICIENT TORQUE IS AVAILABLE TO ACCELERATE THE MOTOR.

Fig 78
Typical Output—Generator Motor Starting Program

TIME (SECONDS)	GENERATOR VOLTS(P.U.)	EXCITER FIELD VOLTS(P.U.)	MOTOR VOLTS(P.U.)
0-	1.00	1.61	0.94
0.0	0.93	1.61	0.79
0.1	0.92	2.12	0.78
0.2	0.92	2.62	0.78
0.3	0.92	3.13	0.78
0.4	0.94	3.60	0.80
0.5	0.97	3.60	0.83
0.6	0.99	3.60	0.85
0.7	1.00	3.60	0.86

Fig 79
Typical Output—Generator Motor Starting Program

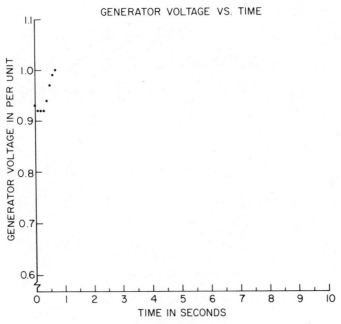

Fig 80
Typical Output — Plot of
Generator Voltage Dip

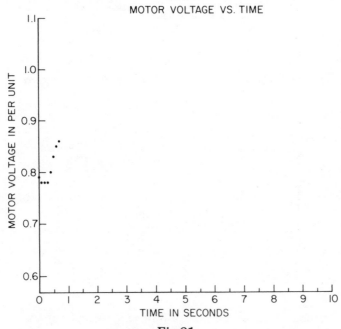

Fig 81
Typical Output — Plot of
Motor Voltage Dip

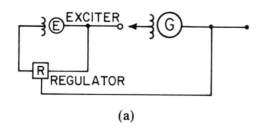

(a)

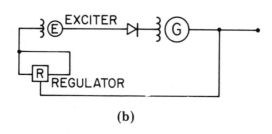

(b)

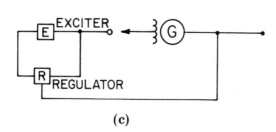

(c)

**Fig 82
Simplified Representation of
Typical Regulator/Exciter
Models for Use in
Computer Programs**

to obtain this report, the excitation system models available are similar to those described in Reference [13]. Excitation system models are shown in simplified form in Fig 82. Continuously acting regulators of modern design permit full field forcing for minor voltage variations (as little as 0.5%), and these voltage changes have been modeled linearly for simplicity.

Variations in exciter field voltage (EFV) over each time interval considered are used to calculate system bus voltages at the end of these same intervals. A single main machine field circuit time constant is used in the generator representation, and Fromlich's approximation [14] for saturation effects is used when the voltage behind the generator leakage reactance indicates that saturation has been reached.

The tabulated output stops just short of full recovery since a more complex model is necessary to represent overshoot, oscillation, etc, beyond this point. Of primary concern is this type of study is the maximum voltage dip and the length of time to voltage recovery as a function of generator behavior and voltage regulator performance.

8.6.5 The Speed-Torque and Motor Accelerating Time Analysis. A simplified sample problem is presented for solution by hand. In this way it is possible to appreciate how the digital computer aids in solving the more complex problems. The following information applies to the system shown in Fig 83.

**Fig 83
Simplified System
Model for Accelerating
Time and Speed-Torque
Calculations**

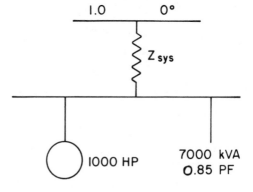

Table 13
Average Values for Accelerating Torque
Over Time Interval Defined by a
Speed Change

speed	T_{motor}	T_{load}	T_{net}	T_{net}
0%	100%	30%	—	—
—	—	—	77.5%	2260.4 lb-ft^2
25%	120%	35%	—	—
—	—	—	100%	2916.7 lb-ft^2
50%	160%	45%	—	—
—	—	—	120%	2500.0 lb-ft^2
75%	190%	65%	—	—
—	—	—	62.5%	1822.9 lb-ft^2
95%	80%	80%	—	—

(1) Motor hp = 1000 (induction)
(2) Motor r/min = 1800
(3) Motor WK^2 = 270 lb-ft^2
(4) load WK^2 = 810 lb-ft^2

Assuming Fig 68 describes the speed-torque characteristic of the motor and the load, it is possible to find an average value for accelerating torque over the time interval defined by each speed change. This can be done graphically for hand calculations, and the results are tabulated in Table 13.

Now applying the simplified formula for starting time provided earlier:

$$t_{0\text{-}25} = \frac{(270 + 810)\,(450 - 0)}{(308)\,(2260.4)} = 0.6981 \text{ s}$$

$$t_{25\text{-}50} = \frac{(1080)\,(900 - 450)}{(308)\,(2916.7)} = 0.5410 \text{ s}$$

$$t_{50\text{-}75} = \frac{(1080)\,(1350 - 900)}{(308)\,(3500.0)} = 0.4580 \text{ s}$$

$$t_{75\text{-}95} = \frac{(1080)\,(1710 - 1350)}{(308)\,(1822.9)} = 0.6925 \text{ s}$$

and therefore, the total time to 95% of synchronous speed (or total starting time) is approximately 2.38 seconds. It can be

seen how a similar technique can be applied to the speed-torque starting characteristic of a wound rotor motor (see Fig 67) to determine the required time interval for each step of rotor starting resistance. The results of such an investigation can then be used to specify and set timers that operate resistor switching contactors or program the control of a liquid rheostat.

The current drawing during various starting intervals can be obtained from a speed-current curve such as the typical one shown in Fig 84. This example has

Fig 84
Typical Motor Speed-Current
Characteristic

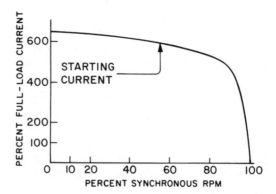

160

MO 1 MOTOR TYPE- 1 TITLE-   BENCH MARK PROBLEM

MO 2 HP- 1000.   SPEED-1775 NEMA TYPE-A SYSTEM VOLTS- 4160.   RATED V- 4000.

MO 3 R1-   A1-   R2-   X2-   X0-

MP 4 NO. PHASES- 3 LOCKED ROTOR CURRENT-   FULL LOAD CURRENT- 145.
* WARNING --- LOCKED ROTOR CURRENT NOT INPUT OR IS INVALID. EXECUTION CONTINUING.

MO 5 INRUSH MULTIPLIER- 6.0 STARTING PF- .15   MOTOR INERTIA WKSQ- 500.

MO 6 MOTOR STARTER DATA--- TYPE- 2 HST-   AST-   TAP

ST 7 NO. POINTS-
* WARNING --- PROGRAM ASSUMING TYPICAL NEMA TYPE A TORQUE-SPEED CURVE.

SC 8 NO. POINTS-
* WARNING --- PROGRAM ASSUMING A TYPICAL SPEED-CURRENT CURVE.

PF 9 NO. POINTS-
* WARNING --- PROGRAM ASSUMING A TYPICAL SPEED-POWER FACTOR CURVE.

LD10 NO. POINTS-   LOAD TYPE-   NO LOAD
* WARNING --- LOAD'S TORQUE-SPEED CURVE MUST BE INPUT, OR IS ASSUMED AS NO LOAD CURVE. EXECUTION CONTINUING.

LD11 LOAD INERTIA WKSQ-   GEARING RATIO-
* WARNING --- MOTOR TO LOAD GEAR RATIO NOT INPUT OR IS INVALID. PROGRAM ASSUMES -- 1. EXECUTION CONTINUING.

SD12 SYSTEM IMPEDANCE HS-   .00101 XS-   .0101

SD13 SYSTEM GENERATION-   IMPEDANCE RG-   XG-

SD14 INITIAL SYSTEM LOAD TYPE- 3 RL-   XL- 7000.   PF- .85

Fig 85
Typical Output—Motor Speed-Torque and Accelerating Time Program

assumed full voltage available to the motor terminals, which is an inaccurate assumption in most cases. Actual voltage available can be calculated at each time interval. The accelerating torque will then change by the square of the calculated voltage. This process can be performed by graphically plotting a reduced voltage speed-torque curve proportional to the voltage calculated at each time interval, but this becomes tedious in a hand calculation. Sometimes, in the interest of simplicity, a torque corresponding to the motor terminal voltage at the instant of the maximum voltage dip is used throughout the starting interval. More accurate results are possible with digital computer program analysis. A sample output report for the analysis is shown in Fig 85.

8.7 Summary. Several methods for analyzing motor starting problems have been presented. Types of motor starting studies available range from simple voltage drop calculation to the more sophisticated motor speed-torque and acceleration time study which approaches a transient stability analysis in complexity. Each study has an appropriate use and the selection of the correct study is as important a step in the solution process as the actual performance of the study itself. Examples presented here should serve as a guide for when to use each type of motor starting study, what to expect in the way of results, and how these results can be beneficially applied. The examples should also prove useful in gathering the required information for the specific type of study chosen. Experienced consulting engineers and equipment manufacturers can give valuable advice, information, and direction regarding the application of motor starting studies as well.

8.8 References

[1] HARBAUGH and PONSTINGL, How to Design a Capacitor Starting System for Large Induction and Synchronous Motors, IEEE IAS 1975 Annual Meeting — Conference Record.

[2] *Industrial Power Systems Handbook*, D. BEEMAN, Editor, New York, McGraw-Hill, 1955.

[3] *Electrical Transmission and Distribution Reference Book*, Westinghouse Electric Corporation, East Pittsburgh, Pennsylvania, 1964.

[4] ANSI/IEEE C37.010-1979, IEEE Application Guide for AC High Voltage Circuit Breakers Rated on a Symmetrical Current Basis.

[5] CROFT, T., CARE, C., and WATT, J. *American Electrician's Handbook*, New York, McGraw-Hill, 1970.

[6] ANSI/NFPA Std 70-1975, National Electrical Code 1975, Boston, Mass.

[7] MANNING, L. *Electrical Circuits*, New York, McGraw-Hill, 1966.

[8] NEUENSWANDER, J. *Modern Power Systems*, International, 1971.

[9] STAGG and EL-ABIAD *Computer Methods in Power System Analysis*, New York, McGraw-Hill, 1971.

[10] WEIDMER, R., and SELLS, R. *Elementary Classical Physics*, 2 vol, Boston, Allyn and Bacon, Inc, 1965.

[11] FITZGERALD, A., KINGSLEY, C., and KUSKO, A. *Electric Machinery*, New York, McGraw-Hill, 1971.

[12] PETERSON, H. *Transients in Power Systems*, Dover, New York, 1951.

[13] IEEE POWER GENERATION COMMITTEE *Computer Representation*

of Excitation Systems, Paper 31 TP 67-424, May 1, 1967.

[14] KIMBARK, E. *Power System Stability: Synchronous Machines*, Dover, New York, 1956.

[15] NEMA Standards Publication ICS 1-1978, General Standards for Industrial Control and Systems; ICS 2-1978, Industrial Control Devices, Controllers and Assemblies; ICS 3-1978, Industrial Systems; ICS 4-1977, Terminal Blocks for Industrial Control Equipment and Systems; ICS 5-1978, Resistance Welding Control; ICS 6-1978, Enclosures for Industrial Controls and Systems, Washington DC.

[16] NEMA Standards Publication ICS-1975, *Industrial Controls and Systems*, New York, New York.

9. Harmonic Analysis Studies

9.1 Introduction. This section discusses the basic concepts involved in studies of harmonic analysis of industrial and commercial power systems. The need for such analysis, recognition of potential problems, corrective measures, required data and benefits derived from performance of a harmonic analysis study are also discussed. Benefits of using a computer as a tool for a harmonic analysis study will also be addressed within this section.

The prevailing sources of harmonics in a system are rectifiers, dc motor drives (converters/inverters), uninterruptible power supplies (UPS), cycloconverters, arc furnaces, or any other device with nonlinear characteristics, which derive their power from a linear/sinusoidal electric system. Systems composed of these types of loads have the potential to develop harmonic related problems and are therefore prime candidates for a harmonic analysis study. Fractional horsepower drives and other single phase loads or small three phase loads fitting into the above categories normally have minimal effect on the system harmonic content and therefore are neglected.

9.2 History. For years motors and other loads requiring dc power derived their energy from ac motor driven dc generators (MG sets). Mechanical linkage between the two systems transmitted power between them and at the same time electrically isolated each system from the other. However, these MG sets were bulky and tended to be high maintenance pieces of equipment.

The first attempt at electrical rectification was accomplished through mechanical means. A motor driven cam physically opened and closed switches at precisely the right instant on the voltage waveform to supply dc voltage and current to the load. At best this approach was cumbersome since timing the switches and keeping them timed was extremely difficult. In addition, contact arcing plus mechanical wear also made this equipment a high maintenance item. Mechanical rectifiers were soon replaced by static equipment including mercury, selenium, and silicon diodes, and finally thyristors (SCRs).

Although solid-state rectification appeared to be the panacea to the problems of the older methods, other system problems soon became noticeable especially as the size of each converter unit and as the total converter load became a substantial section of the total system power requirements.

The most noticeable initial problem was the inherent poor power factor associated with static rectifiers. Economics as well as system voltage regulation requirements made it desirable to improve the overall system power factor which normally was accomplished using shunt power factor correction capacitors. However, when these capacitor banks were applied, other problems involving harmonic voltages and currents affecting these capacitors and other related equipment became prevalent.

Another initial problem was the excessive amount of interference induced into telephone circuits due to mutual coupling between the electrical system and the communication system at these harmonic frequencies.

More recent problems involve the performance of computers, numerical controlled machines and other sophisticated electronic equipment which are very sensitive to power line pollution. These devices can respond incorrectly to normal inputs, give false signals, or possibly not respond at all.

9.3 General Theory.

9.3.1 What are Harmonics? Harmonics are voltages or currents, or both, present on an electrical system at some multiple

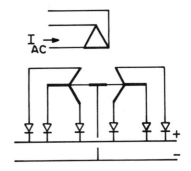

Fig 86
6-Phase, 6-Pulse
Rectifier

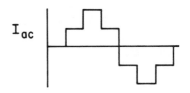

Fig 87
6-Phase, 6-Pulse
Rectifier

of the fundamental (normally 60 Hz) frequency. Typical values are the 5th (300 Hz), 7th (420 Hz), 11th (660 Hz) and so on.

To better understand harmonic related problems, it is necessary to understand how and where harmonics are generated.

In converting ac power to dc power, a rectifier effectively breaks or chops the alternating current waveform by allowing the current to flow only during a section of the cycle. The example in Figs 86 and 87 for a 6-phase, 6-pulse rectifier indicates the waveforms for the direct current and the corresponding ac line current. The square alternating current waveform represents a distorted sinusoidal waveform rich in harmonic content which can be resolved into components using Fourier analysis techniques [1]. The Fourier series for this waveform is:

$$I_{ac} = \frac{2\sqrt{3}}{\pi} I_d \left(\cos\theta - \frac{1}{5}\cos 5\theta \right.$$
$$+ \frac{1}{7}\cos 7\theta - \frac{1}{11}\cos 11\theta$$
$$\left. + \frac{1}{13}\cos 13\theta \ldots \right)$$

The higher frequency terms are the harmonic components.

A similar Fourier analysis of the distorted sinusoidal waveforms of other harmonic generating equipment as mentioned previously will yield similar harmonic current components.

Arc furnaces differ from drives and rectifiers in that harmonic voltages are generated instead of harmonic currents. Arc resistance and the voltage/current characteristis are continually varying due to movement of scrap, bubbling of molten metal, magnetic repulsion of the arc from the other two phases, and so forth. In addition, magnetic repulsion forces between furnace flexible cables cause swinging of these cables resulting in variation of the reactance of secondary circuit. The overall result of the non-linearity of the arc and arc furnace parameters is the generation of harmonic voltages in the secondary circuit. Because of the unpredictable nature of the arc, harmonic magnitudes are not readily determined. Only lower order harmonics are generated, however.

With harmonic sources present, all that remains is to have a path in the ac system for these harmonics to flow. If this occurs, harmonic related problems can have a detrimental effect on the ac system.

Various parameters particular to each system determine the magnitude of these harmonic problems.

9.3.2 Resonance. The application of capacitors with harmonic generating apparatus on a power system necessitates the consideration of the potential problem of an excited harmonic resonance condition.

Inductive reactance increases directly with frequency and capacitive reactance decreases directly with frequency. At the resonant frequency of any inductive-capacitive (LC) circuit, inductive reactance equals the capacitive reactance.

There are two forms of resonance to be considered, series resonance and parallel resonance. For the series circuit in Fig 88, the total impedance at the resonant frequency reduces to the resistance component only. For the case where this component is small, high current magnitudes at the exciting frequency will flow.

Figure 89 is a plot of impedance versus frequency of this series circuit.

Fig 88
Series Circuit

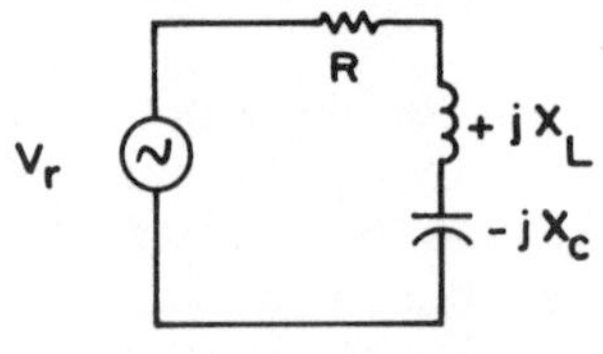

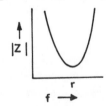

Fig 89
Impedance Versus Frequency

Parallel resonance is similar to series resonance in that at its exciting frequency, capacitance reactance equals the inductive reactance. However, its parallel impedance is significantly different. Figure 91 is a plot of impedance versus frequency for the parallel circuit of Fig 90. At the resonant frequency f_r, the impedance is very high and when excited from a source at this frequency, a high circulating current will flow in the capacitance-inductance loop although the source current is small in comparison.

To illustrate parallel resonance further, select 60 Hz reactances of 0.60 Ω and -30.23 Ω for X_L and X_C respectively. The parallel impedance of the capacitor and inductor at 60 Hz can be calculated to +0.61 Ω. For illustration purposes, inject 1 A (60 Hz) current I_f into the circuit. Using Ohm's law, the voltage across the load V_L due to the fundamental frequency is then 0.61 V.

Fig 90
Parallel Circuit

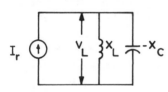

Fig 91
Impedance Versus Frequency

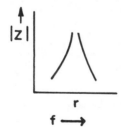

For this same circuit assume a current source equal in magnitude to $1/7$ A $I_f/7$ at the 7th harmonic (420 Hz). At this frequency the inductive reactance becomes 4.2 Ω and the capacitive reactance becomes -4.4 Ω. The parallel combination can be calculated to be 92.4 Ω. Again, using Ohm's law, the voltage across the load (V_L) now is 13.20 V with the current through the capacitor being 3.0 A.

In actual electrical systems utilizing power factor correction capacitors, either type of resonance or a combination of both can occur if the resonant point happens to be close to one of the frequencies generated by harmonic sources in the system. The result can be the flow of excessive amounts of harmonic current or the appearance of excessive harmonic overvoltages, or both. Possible consequences of such an occurrence are excessive capacitor fuse operation, capacitor failures, telephone interference, or overheating of other electrical equipment.

9.4 Modeling. To analyze a system for resonance effects requires calculation of various harmonic currents throughout the system and the harmonic voltages these currents cause. At each frequency, system impedances are different. For all conditions the circuit remains the same. Rectifiers and other similar harmonic generating equipment are represented as current sources at each harmonic frequency. With reference to the previous Fourier expansion, the maximum theoretical harmonic current magnitude from each converter equals the fundamental frequency full load current magnitude divided by the order of the harmonic.

Harmonic current magnitudes are also functions of converter pulses. Magnitudes of system harmonic voltages are a result of harmonic currents flowing back into the harmonic impedances of the ac system. The order of harmonic currents is $np \pm 1$, where n is any integer and p is the number of rectifier pulses. Thus, for a 6-pulse rectifier, the order of harmonics are 5th, 7th, 11th, 13th, 17th, 19th etc. For a 12-pulse rectifier, the order of harmonics are 11th, 13th, 23rd, 25th, 35th, 37th, etc. This procedure using a higher number of phases for lower order harmonic cancellation is referred to as phase multiplication. Although phase multiplication theoretically will cancel normal harmonics not of the order $np \pm 1$, in practice, both current magnitude and phase angles deviate enough to allow only incomplete cancellation. Most literature on the subject indicates that 10% to 25% of the maximum harmonic magnitude will remain [2]. To be as realistic as possible, this factor, sometimes referred to as harmonic cancellation factor (HCF), should be included.

Additional reduction of the harmonic current magnitude is due to the series inductive reactance between the harmonic source and the utility supply. The larger this inductive reactance is (commutating reactance) the more it impedes that particular harmonic generation. For example, the maximum 5th harmonic current magnitude available is $1/h = 1/5 = 0.2$ per unit or 20% (commutating reactance = 0). Due to a significant commutating reactance, however, the actual magnitude is only 17%. This reduction is referred to as commutating reactance factor (CRF). Graphs of per unit I_h (CRF) versus commutating reactance are available in the appropriate standards [3].

The final factor reducing a particular harmonic current is the per unit loading (LDF). If a rectifier is only 50% loaded (fundamental component of current) then the harmonic current will only be 50% of its maximum value on that sys-

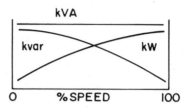

Fig 92
Typical Thyristor Drive
Characteristics

tem. For drives with phase retard control, this loading factor also is proportional to the ac fundamental current component but is not necessarily proportional to the dc, kW, or hp output.

The graph in Fig 92 is typical for thyristor drives with the kVA value approximately constant over the entire range of speeds. However, the loading factor in question is the fundamental component only and for low speeds, the kvar component of the kVA term is very rich in harmonic content. The assumption of constant kVA and maximum loading at maximum phase retard (minimum speed) yields conservative results.

The total harmonic current value injected into the system by a particular device is then: $I_h = (I_{FL}/h)\,(LDF)\,(HCF)\,(CRF)$ where I_{FL} is the fundamental full load current of the device and h is the order of harmonic.

9.4.1 Analysis Techniques. With basic system connections (one-line diagram) and impedances established, a harmonic analysis study should analyze the system under steady-state conditions for normal power flow and harmonic current flow (sometimes referred to as harmonic load flow) for all the harmonics being modeled for as many system switching conditions as required. A typical range of harmonic frequencies modeled can be from 5 (300 Hz) to 37 (2220 Hz). The harmonic resonant point at a particular location will probably differ under each separate switching condition so all normal modes of operation should be included.

Since each switching condition requires many solved cases of the system for each of the harmonics being modeled, the digital computer solves the multitude of calculations required for each switching condition solution. The time required for all but the simplest systems normally prohibits manual calculations.

Power factor correction capacitors are designed to continuously carry 135% of their nameplate rated (fundamental) kVA, or kV_{ac}, 110% of their rated voltage and 180% of their rated current in the existing standard (135% current in the pending standard). This *overload* capability allows a margin for system overvoltages or harmonic voltages, or both, which can occur. The total loading of a bank is calculated as the sum of the kVA loadings of the fundamental and each harmonic.

This is expressed as

$$kV_{ac} = \sum_h (V_h I_h) = \sum_h (V_h^2 / X_h)$$

where

> h = order of harmonic (including the first or fundamental)

Detailed results of a computerized harmonic analysis study should include the four main points mentioned above: total capacitor bank kV_{ac} loading, peak voltages, rms current, and rms voltage. Voltage and current values should be provided at all critical system locations susceptible to harmonic problems where appropriate values can be compared to the device's ratings in question.

$$V_{peak} = \sum_{h=1} V_h$$

The peak is used because of the *random* phase relationships existing between the various harmonic components. The third loading factor is the total rms current which connections, bushings, and other components of the capacitor bank must handle. The current is calculated as:

$$I_{rms} = \left(\sum_h I_h^2 \right)^{1/2}$$

Additional problems arise from harmonics in motors, lighting ballast transformers and other similar equipment. These problems are essentially excessive heating due to circulating harmonic currents. To evaluate this effect, rms voltages rather than peak values are required. Therefore, rms voltages should also be calculated and printed throughout the system. Total rms voltage can be calculated using the following equation.

$$V_{rms} = \left(\sum_h V_h^2 \right)^{1/2}$$

where

> h = order of harmonic (including the first or fundamental)
> V = fundamental or harmonic voltage
> I = fundamental or harmonic current
> X = fundamental or harmonic reactance

Capacitors must also have sufficient dielectric to withstand anticipated peak voltages resulting from the fundamental and the harmonics. This peak voltage is calculated as the arithmetic sum of all the component voltages (not as rms value).

9.5 Solutions to Harmonic Problems. The primary solution to any harmonic related problem is accomplished by shifting the system resonant point to some other frequency not generated by the electrical equipment of the system. The simplest and least expensive method is to alter or bypass system operating conditions and procedures that lead to harmonic resonance.

If this approach is impractical or undesirable then quite often additional apparatus is required.

Remedial measures involving additional equipment generally used to minimize harmonic effects include shunt L-C filters located at the harmonic source and tuned to series resonance at the troublesome harmonics. This approach makes a low impedance path for the harmonic currents to flow with very little flowing back into the rest of the ac system. However, a separate filter is required for every major harmonic source.

In other cases where power factor correction capacitors in the ac system cause resonance at the generated harmonics, their location or size can be changed to eliminate the resonance, or series reactors can be added to detune them at the troublesome resonant frequency.

There are three different major schemes which will accomplish adequate filtering. Economic considerations as well as the particular filtering requirements for each case determine which scheme is most desirable. Utility requirements for harmonic content injected into its system dictate which scheme is to be used. Figs 93, 94 and 95 indicate these schemes.

The least expensive and therefore most desirable of these three schemes is that shown in Fig 93. Generally, a carefully selected tuning reactor is sufficient to alleviate harmonic resonance. Careful selection of the tuning reactor is stressed. If incorrectly selected, the harmonic frequency for which it is tuned will probably be lowered to acceptable levels but

Fig 93
No 1 Scheme for
Adequate Filtering

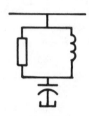

Fig 94
No 2 Scheme for
Adequate Filtering

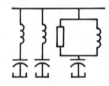

Fig 95
No 3 Scheme for
Adequate Filtering

another harmonic frequency can then become dominant with the resonant point only being shifted to another harmonic frequency present on the system.

Figures 94 and 95 are used when more stringent harmonic content requirements are in effect. In Fig 95 the two lowest and most troublesome harmonics are filtered out individually with one high pass filter.

For new installations or major expansions on existing facilities which have a significant amount of rectification there is another method of reducing lower order harmonics if a harmonic analysis is incorporated within the design stages. The concept of phase multiplication mentioned previously and the inherent phase cancellation of lower order harmonics can be employed to greatly reduce harmonic magnitudes. From an economic and manufacturing viewpoint, it is impractical to construct rectifier transformers with more than twelve phases. However, they can be selected to collectively appear as more phases than they actually are.

In Fig 96, both rectifier transformers are individually 6-phase units but when viewed from bus A and when they are both equally loaded, they collectively appear to be a 12-phase system. Similarly in Fig 97, buses C and D appear to have 12-phase rectification but due to differing connections of the power transformers, the system becomes a 24-phase system at bus B.

Other methods of accomplishing phase multiplication entail different phase

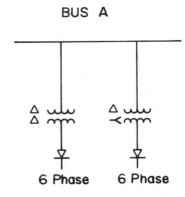

Fig 96
6-Phase Rectifier
Transformers

shifting of the rectifier transformers themselves. For example, one rectifier rated as 12-phase but with six phases shifted $-30°$ and six phases at $0°$ with respect to the primary system will appear to be a 24-phase system when operated with another identically rated and loaded 12-phase rectifier which has $-15°$ and $+15°$ phase shifts.

It must be stressed that lower order

Fig 97
24-Phase System

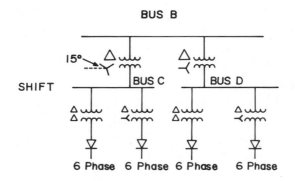

harmonic cancellation using phase multiplication techniques is only applicable when each component is equally rated and loaded. Where not all units are equally loaded or some are off line only partial lower order harmonic cancellation occurs. A harmonic analysis study should also include probable imbalanced loading conditions with only partial lower order harmonic cancellation considerations.

Often an installation using phase multiplication for harmonic reduction might not need any filtering at all or requires only small tuning reactors opposed to a complex RLC filtering scheme. The net economic effect is significant.

Examples: The partial one-line diagram in Fig 98 was taken from an actual harmonic analysis study performed in the design stages for an electro-chemical plant.

This particular system required the addition of tuning reactors in series with each capacitor bank to minimize harmonic resonance problems.

Tables 14 and 15 are abbreviated copies of the computer solutions for this system. The first solution modeled the system without any filters to determine if a resonant condition would exist after application of the proposed capacitor banks. As indicated by the harmonic profile at C3 bus (2 banks), the system is resonant very close to the 7th harmonic. A comparison of capacitor bank ratings to the values listed in the computer printout indicated that filtering was indeed required.

The application of tuning reactors R3A and R3B tuned to the 4.7th harmonic was sufficient to suppress this resonant condition to acceptable levels. The second computer solution indicates the 7th harmonic to be almost gone with the totalized quantities greatly reduced. The rms voltage was lowered from 1.766 per unit to 1.102 per unit of the system base value. The current and reactive power loads were well within ratings. The voltage ratings of the proposed banks were increased to 109% of the nominal

Fig 98
Partial One-Line Diagram

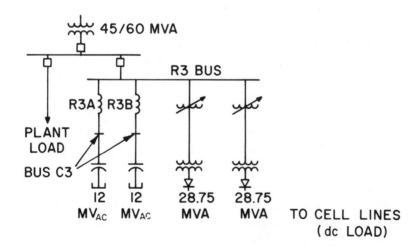

172

Table 14
First Computer Solution:
Without Filters

WESTINGHOUSE ELECTRIC CORPORATION - HARMONIC ANALYSIS STUDY

SOLUTION FOR SWITCHING CONDITION 195

BUS NAMES	C3	C2	C1	R3	R2	R1

5TH HARMONIC BUS VOLTAGES AND SHUNT LOAD CURRENTS

	C3	C2	C1	R3	R2	R1
BUS VOLTAGES	.135	.009	.009	.135	.009	.009
LOAD CURRENT	.673	.044	.043	.673	.044	.043

7TH HARMONIC BUS VOLTAGES AND SHUNT LOAD CURRENTS

	C3	C2	C1	R3	R2	R1
BUS VOLTAGES	1.414	.115	.117	1.414	.115	.117
LOAD CURRENT	9.896	.805	.819	9.896	.805	.819

11TH HARMONIC BUS VOLTAGES AND SHUNT LOAD CURRENTS

	C3	C2	C1	R3	R2	R1
BUS VOLTAGES	.009	.004	.004	.009	.004	.004
LOAD CURRENT	.101	.048	.046	.101	.048	.046

13TH HARMONIC BUS VOLTAGES AND SHUNT LOAD CURRENTS

	C3	C2	C1	R3	R2	R1
BUS VOLTAGES	.004	.002	.002	.004	.002	.002
LOAD CURRENT	.055	.030	.028	.055	.030	.028

17TH HARMONIC BUS VOLTAGES AND SHUNT LOAD CURRENTS

	C3	C2	C1	R3	R2	R1
BUS VOLTAGES	.002	.000	.000	.002	.000	.000
LOAD CURRENT	.035	.008	.008	.035	.008	.008

19TH HARMONIC BUS VOLTAGES AND SHUNT LOAD CURRENTS

	C3	C2	C1	R3	R2	R1
BUS VOLTAGES	.001	.000	.000	.001	.000	.000
LOAD CURRENT	.023	.005	.005	.023	.005	.005

TOTALIZED QUANTITIES INCLUDING ALL ABOVE HARMONICS PLUS FUNDAMENTAL

	C3	C2	C1	R3	R2	R1
BUS V(ARITH)	2.617	1.183	1.184	2.617	1.183	1.184
BUS V(RMS)	1.766	1.056	1.057	1.766	1.056	1.057
LOAD I(RMS)	9.975	1.325	1.333	9.975	1.325	1.333
LOAD KVA	15.184	1.196	1.199	.000	.000	.000

173

Table 15
Second Computer Solution:
With Filters

WESTINGHOUSE ELECTRIC CORPORATION - HARMONIC ANALYSIS STUDY

SOLUTION FOR SWITCHING CONDITION195

BUS NAMES	C3	C2	C1	R3	R2	R1

5TH HARMONIC BUS VOLTAGES AND SHUNT LOAD CURRENTS

	C3	C2	C1	R3	R2	R1
BUS VOLTAGES	.062	.020	.020	.008	.003	.003
LOAD CURRENT	.294	.093	.096	.294	.093	.096

7TH HARMONIC BUS VOLTAGES AND SHUNT LOAD CURRENTS

	C3	C2	C1	R3	R2	R1
BUS VOLTAGES	.014	.006	.006	.017	.008	.008
LOAD CURRENT	.094	.043	.043	.094	.043	.043

11TH HARMONIC BUS VOLTAGES AND SHUNT LOAD CURRENTS

	C3	C2	C1	R3	R2	R1
BUS VOLTAGES	.002	.005	.005	.010	.012	.012
LOAD CURRENT	.023	.049	.049	.023	.049	.049

13TH HARMONIC BUS VOLTAGES AND SHUNT LOAD CURRENTS

	C3	C2	C1	R3	R2	R1
BUS VOLTAGES	.001	.003	.003	.008	.009	.010
LOAD CURRENT	.014	.036	.036	.014	.036	.036

17TH HARMONIC BUS VOLTAGES AND SHUNT LOAD CURRENTS

	C3	C2	C1	R3	R2	R1
BUS VOLTAGES	.001	.000	.000	.008	.004	.004
LOAD CURRENT	.010	.005	.005	.010	.005	.005

19TH HARMONIC BUS VOLTAGES AND SHUNT LOAD CURRENTS

	C3	C2	C1	R3	R2	R1
BUS VOLTAGES	.000	.000	.000	.006	.003	.003
LOAD CURRENT	.007	.004	.004	.007	.004	.004

TOTALIZED QUANTITIES INCLUDING ALL ABOVE HARMONICS PLUS FUNDAMENTAL

	C3	C2	C1	R3	R2	R1
BUS V(ARITH)	1.180	1.137	1.137	1.131	1.120	1.121
BUS V(RMS)	1.102	1.100	1.100	1.050	1.050	1.050
LOAD I(RMS)	1.095	1.057	1.058	1.095	1.057	1.058
LOAD KVA	1.174	1.157	1.157	.076	.057	.057

system voltage, bringing all ratings safely within limits.

The values listed under bus R3 are those of the main bus. Very satisfactory system harmonic filtering is also achieved.

The previous example illustrates how effective filtering can be when properly engineered. However, filtering is not always the best solution. In another electrochemical plant with a very large complex electrical system, harmonic filtering was not implemented. The existing system utilized power factor correction capacitors at each major rectifier location. Plant expansion required an additional rectifier and capacitor bank. Initial filter design simulation corrected harmonic problems for the new bank but shifted the resonant point to another location which caused harmonic problems with existing capacitor banks. The practical and economical solution in this case was to oversize the new capacitor bank's voltage rating to compensate for harmonic content without filtering.

9.6 When is a Harmonic Study Required?

Although a specific answer is not always available, the following points are indicators.

(1) Application of capacitor banks to systems comprised of 20% or more of converters or other harmonic generating equipment

(2) History of harmonic related problems including excessive capacitor fuse operation

(3) In the design stage of a facility composed of capacitor banks and harmonic generating equipment

(4) Strict electric power company requirements which limit harmonic injection back into its system to very small magnitudes

(5) Plant expansions which add significant harmonic generating equipment

operating in conjunction with capacitor banks

Occasionally, when harmonics appear to be the cause of system problems, it is desirable to determine the system harmonic resonance point. To determine this resonance point, the short circuit capacity at each capacitor bank location is required. A close approximation of this resonance point is the equation:

$$h_r = \sqrt{\frac{MVA}{MV_{ac}}}$$

where

h_r = resonance point in per unit of fundamental frequency

MVA = short circuit capacity

MV_{ac} = Mvar rating of the unfiltered capacitor bank at that location [2]

This equation is useful for an initial evaluation. If the resonance point is close to one of the harmonic frequencies present on the system, then possible harmonic related problems could occur.

9.6.1 Data Required. The following data are required for a typical study:

(1) Single-line interconnection diagram

(2) Short circuit capacity and X/R ratio of the utility supply system

(3) Subtransient reactance and kVA of all rotating machines. Where possible, all machines on a given bus should be lumped together into one composite equivalent machine

(4) Percent reactance and resistance of all lines, cables, bus work, current limiting reactors, and saturable reactors on a given kVA base and rated kV of the circuit in which the circuit element is located

(5) The percent impedance and kVA of all power transformers

(6) The three-phase kvar rating of all shunt capacitors and shunt reactors

(7) Namplate ratings, number of phases, phase connections, whether diodes or thyristors and if thyristors, the maximum phase delay angle, per unit loading and loading cycle of each rectifier unit connected to the system. Actual manufacturers' test sheets on each transformer are also helpful but not absolutely mandatory

(8) Specific system configurations and operating procedures for the rectifier circuits being studied

(9) Actual maximum normally expected voltage for the system supplying the rectifier loads

(10) For arc furnace installations, secondary impedance from the transformer to the electrode tips plus a loading cycle to include arc MW, secondary voltages, secondary current and furnace transformer taps

9.7 References.

[1] KIMBARK, E. A. *Direct Current Transmission*, New York, John Wiley and Sons, 1971.

[2] STEEPER, D. E., and STRATFORD, R. P. Reactive Compensation and Harmonic Suppression for Industrial Power Systems Using Thyristor Converters, Oct 7–10, 1974.

[3] IEEE Std 444-1973, IEEE Standard Practices and Requirements for Thyristor Converters for Motor Drives.

[4] ADAMSON, C., and HINGORANI, N. G., *High Voltage Direct Current Power Transmission*, London, Garraway, 1960.

[5] BORREBACH, E. J. The Effect of Arc Furnace Loads on Power Systems, IEEE/IAS Conference Record, Paper MI-Thu-Am3 1085, Oct 7–10, 1974.

[6] McFADDEN, R. H., RAMEY, D. G., and SHANKLE, D. F., Electrical Transients in Arc Furnace Installations, IEEE/IAS I and CPS Conference Record, May 2, 1972.

[7] SCHIEMAN, R. G. Electrical Circuit Problems Caused by SCR Drives, IEEE Pulp and Paper Conference, June 17–19, 1970.

[8] SCHIEMAN, R. G., and SCHMIDT, W. C. Power Line Pollution by 3-Phase Thyristor Motor Drives, IEEE/IAS Annual Meeting, 1976.

10. Switching Transient Studies

10.1 Introduction. Compared to classical steady-state power system studies (for example, short-circuit, load flow, protective device coordination, etc) switching transient investigations in depth are conducted quite infrequently in industrial power systems, with arc furnace melt shop systems accounting for most such studies. The relatively high rate of switching that occurs in arc furnace systems, the prevalence of capacitors, and the erratic nature of the arc furnace load, all lead toward a greater need to understand the associated transient duties. Relatively superficial investigations of limited scope are commonly undertaken, however, in industrial systems, to assist resolution of certain transient behavioral questions in conjunction with the application or failure of a particular piece of equipment.

Two basic approaches present themselves in the determination and prediction of switching transient duties in a power system. They can be determined by direct measurements and monitoring or analytically by calculation/modeling techniques. This section gives an orientation of the basic analytical techniques in general use, their respective areas of applicability, and the associated data and equipment requirements; salient considerations in planning and conducting field measurements are also presented. The most common approach to the analytical determination of transient duties in industrial power systems, the transient network analyzer, is elaborated upon, including excerpts from an actual industrial transient study.

10.2 Basic Concept of Switching Transients. The switching event in a power system initiates the transition between two steady-state conditions, the pre-switching condition and the post-switching condition. In steady state, the energy stored in various inductances and capacitances of the dc circuit are constant. In an ac circuit, energy is continually exchanged cyclically between circuit inductances and capacitances. Depending upon resistances present, losses will extract energy which will be supplied by various sources within the system. Each steady-state condition entails its own unique set of energy storage and exchange rates in and among the various circuit elements.

Thus a redistribution of energy must occur among the various system elements to change from one steady-state condition to another. This change cannot occur instantly; a finite period of time, the transient period, prevails during which transient voltages and currents develop to bring about these changes. These transient voltages and currents develop and proceed in an orderly manner, prescribed by the network configuration and conditions prevailing before and after the switching event. The precise nature and timing of the switching event itself will profoundly affect the characteristics of the ensuing transients. These transient voltages and currents are composed of damped natural-frequency oscillations which potentially can magnify to many times normal. The transient environment is based on Kirchhoff's law. Since changing quantities and their associated rates of change are involved throughout the transient period, the mathematics describing transients involves differential equations. Classical mathematical computation of electrical transients, therefore, requires a certain degree of mathematical proficiency and effort.

In the broad sense, virtually all electrical transients are switching-produced transients if switching is anything that suddenly changes the network configuration or any of its elements. Switching, for example, is considered to be not only the intentional actions of opening and closing circuit breakers and switches, but also the occasions of fault inception,

restriking in switches and breakers, intermittent or arcing faults, clearing by fuse blowing, and so forth. While in this broad sense lightning is also a switching-produced transient, it is usually listed in a class by itself.

10.3 Control of Switching Transients. The philosophy of mitigation and control of switching transients revolves around:

(1) Minimizing the number and severity of switching events

(2) Limitation of the *rate* of exchange of energy that prevails among system elements during the transient period

(3) Extraction of energy

(4) Shifting of resonance points to avoid amplification of particularly offensive frequencies

(5) Provision of energy reservoirs to contain released or trapped energy within safe limits of current and voltage

(6) Provision of preferred paths for elevated-frequency currents attending switching

This philosophy is implemented practically through the judicious use of:

(1) Temporary insertion of resistance between circuit elements, for example, insertion resistors in circuit breakers

(2) Inrush control reactors

(3) Damping resistors in filter circuits and surge protective circuits

(4) Tuning reactors

(5) Surge capacitors

(6) Filters

(7) Surge arresters

(8) Necessary switching only, with properly maintained switching devices

(9) Proper switching sequences

10.4 Methods of Analysis. While very few practicing industrial plant engineers are proficient at the type of mathematics used in the description and analysis of transients, a very useful insight into the physical aspects prevailing in the network during the transient period can be obtained with a minimum of mathematical rigor for the simpler circumstances. In fact, practiced transient analysts use known response patterns based on a few basic fundamentals to assess general transient behavior and to judge the validity of complex transient study results. Indeed, simple arrangements composed of linear circuit elements, and, say, involving only one or two switching events, can be processed *by hand*. Beyond these relatively simple limits, however, the economical calculation of power system switching-produced transients is actually done with one or more, or a combination of, basic analysis aids that have been developed over several decades.

10.5 Analysis Aids. Perhaps the earliest aids to analysis of power system transients were so-called mechanical differential analyzers and later the electronic differential analyzer, most commonly called *analog computer*. These devices assist the solution of differential equations that evolve in mathematical representation of transient phenomena.

Early in the application of analog computers it was recognized that many problems involving abnormal voltage conditions could not be solved by conventional methods. Specifically, systems to be studied for transient voltages due to system switching operations, as well as certain steady-state problems involving nonlinear circuit elements such as surge arresters or saturable transformers or reactors, were awkward to represent except by model techniques. This led to the electromagnetic model approach to transient analysis wherein individual electric power system elements are represented by miniature counterparts with nearly identical electric and magnetic

response (including nonlinearities) to harmonics, traveling waves, and switching surges. These model components, composed into a multi-phase analog representation of the actual power system being investigated, are called a transient network analyzer (TNA).

Digital computers are now being used in transient analysis. In limited areas, digital computers are replacing the transient network analyzer (TNA). For simple system arrangements without transformers or other saturating devices, some digital programs now offer an economical alternative to the TNA. For more complex systems arrangements with nonlinear circuit elements, however, progress has been slower but substantial.

Economical analog-to-digital and digital-to-analog conversion techniques have permitted interfacing of digital and analog computers for a substantial degree of optimization of the most desirable features of each. These so-called hybrid computers or simulators have come into use, principally in the last decade, as another potent tool in the analysis of transients in power systems. They derive their attractiveness from a combination of speed (rapid integrating capability) of the analog computer with the accuracy and flexibility (through use of memory capability) of the digital computer.

The model approach of the TNA finds its virtue in the relative ease with which individual components can duplicate their actual power system counterparts, as compared with the difficulty of representing (mathematically) combinations of nonlinear interconnected elements in an analytical solution; especially attractive are the relative low cost and the speed of solution for individual runs on the TNA. Also, the modern TNA has benefited very substantially from augmentation with digital and analog computers. Such augmentation, for example, offers an economical way to obtain a statistical distribution of switching surge magnitudes. Further, TNA modeling has become a refined art and is becoming increasingly enhanced by electronic techniques. Traditionally and presently, most switching transient studies are conducted through TNA.

There are various equipment approaches and methods to aid the calculation of switching-produced transients to some extent these methodoligies utilize overlapping technologies. While the TNA continues to be the most used aid in the calculated determination of transients, the method employed depends upon preferences, experiences, and equipment available to the analysts. The industrial plant user should entrust transient analysis only to experts using the particular equipment of their preference.

10.6 Data Required for a Switching Transient Study. Compared to conventional power system studies, switching transient analysis data requirements are often more detailed and unusual or special. These requirements remain essentially unchanged regardless of basic analysis tools and aids that are employed, be they digital computer, hybrid simulator, or transient network analyzer, or a combination of these. The generalized data listed below encompass virtually all information areas required in an industrial power system switching transient study:

(1) One-line diagram of the system showing all circuit elements and connection options

(2) Utility information, for each tie, at the connection to the plant, exclusive of the plant load or backfeed

(a) Impedances R, X_L, X_C, both positive and zero sequence, representing utility system or systems under present minimum and future maximum short-circuit duty conditions, with MVA and voltage bases if impedances are in per unit

(b) Voltage spread: maximum and minimum voltage limits

(c) Operation: description of reclosing procedures and contractual or other limitations, if any

(3) Individual power transformer data and other transformer data, if any

(a) Rating, connections, no-load tap voltages for both positions of the Y-Δ switch, if any, and LTC voltages, if any. Location of taps, in LV or HV windings. Normal position of the no-load tap if seldom varied.

(b) No-load saturation data: Curve of no-load voltage versus exciting current, additionally specifying rated voltage magnetizing impedance (or exciting current at rated voltage) and air-core impedance. Definition of winding arrangement for which the data apply. Bases should be given for voltages or currents in percent or per unit.

(c) Positive and zero sequence leakage impedances, R and X_L, for all transformer tap connections. Impedance of series reactor, if any

(d) Neutral grounding details

(4) Capacitor data, for each Mvar supply bank and surge protection unit

(a) Mvar or microfarad rating, voltage rating, catalog number, connections, neutral grounding details

(b) Description of switching device for Mvar supply bank

(c) Description of tuning reactors, if any, for Mvar supply bank

(5) Feeder cables or lines: Impedances per phase, R, X_L, X_C, both positive and zero sequence, for each circuit of appreciable length. If these impedances are not available directly, to permit their calculation, data must be given as follows:

(a) Length of circuit, wire size, conductor arrangement and material, for phase conductors and ground conductors, or other return circuit paths. If cable, whether in magnetic or non-magnetic duct, voltage rating, insulation specification, shield and sheath description.

(b) If multiple conductors per phase, a dimensioned cross section of the feeder installation (for example, of a duct bank for multiple under ground cables) showing several conductors of each phase and ground return conductors.

(6) Other power system elements.

(a) Surge arresters: location, rating, catalog or model number

(b) Grounding resistors (and reactors, if any): rating, impedance of each

(c) Buffer reactors: rating, impedance for all taps of each

(d) Rotating machines: rating of each, subtransient and transient reactance, type of regulator.

(7) Operating modes and procedures.

(a) Sequence and occasion for closing each switch and circuit breaker

(b) Action of existing protection scheme during system overvoltages and undervoltages

Transformer data requirements (3) include items which require considerable time for the transformer manufacturer to develop. This should be factored into the lead time in projecting the date of study completion.

10.7 Switching Transient Problem Areas. Switching of predominately reactive equipment represents the greatest potential for creating excessive transient duties. Principal offending situations are switching capacitor banks with inadequate or malfunctioning switching devices and energizing and deenergizing transformers with the same switch deficiencies. Capacitors can store, trap, and suddenly release relatively large quantities of energy. Similarly, highly inductive apparatus possesses energy storage capability which can release large quantities of electromagnetic energy during a rapid current decrease. Since transient voltages and currents arise in conjunction with energy redistribution that occurs during the transient period between steady-state conditions, the greater the energy storage in associated system elements, the greater the transient magnitudes become. This has been confirmed countless times in studies.

Generalized switching transient studies have provided many important criteria to enable system designers to avoid excessive transients in most common circumstances. Criteria for proper system grounding to avoid transient over voltages on the occasion of a ground fault are a prime example. Results of these generalized studies have formed the basis of several IEEE Committee reports on switching surges. There are also several not-to-common potential transient problem areas that are analyzed on an individual basis. The following is a partial list of transient-related problems which can, and have been, analyzed in computer studies:

(1) Energizing and deenergizing transients in arc furnace installations

(2) Ferroresonance

(3) Lightning and switching surge response of motors, generators, transformers, transmission towers, cables, etc

(4) Lightning surges in complex station arrangements and optimum surge arrester location

(5) Surge transfer through transformers

(6) Switching of large-magnitude inductive current

(7) Switching capacitors

(8) Restrike phenomena in dropping lines, cables, and capacitor banks, includ-

ing modification by nonlinear transformer and arrester elements

(9) Neutral instability and reversed phase rotation

(10) Rectifier transients

(11) Voltage flicker

(12) Magnification of switching surges

(13) Energizing and reclosing transients on lines or cables, either open-ended or transformer-terminated

(14) Switching surge reduction by means of controlled closing of circuit breaker or resistor pre-insertion, or both

(15) Statistical distribution of switching surges

(16) Recovery voltages on distribution and transmission systems

(17) Energizing transients on systems using phase-shifting transformers

10.8 Switching Transient Study Objectives. The basic objectives of a switching transient investigation are to identify and quantify transient duties that may arise in a system as a result of intentional and non-intentional switching, and to prescribe economical corrective measures, if necessary. Results of a switching transient study can affect operating procedures as well as the equipment in the system. The following include specific broad objectives, one or more of which are included in a given study:

(1) Identify nature of transient duties (that is, magnitude, duration, and frequency) which can occur for any realistic operational (that is, intentional) switching

(2) Determine if abnormal transient duties are likely to be imposed on equipment by the inception of faults and their removal

(3) Recommend corrective measures to mitigate excessive transients, such as resistor insertion, tuning reactors (for capacitor banks), appropriate system

grounding, and application of surge arresters and surge protective capacitors

(4) Recommendation of alternative operating procedures, if necessary, to minimize transient duties

(5) Document results of study on case-by-case basis in a readily understandable form for those responsible for system design and operation. Such documentation usually includes reproductions of wave shape displays of oscilloscopes or various recorders and reproductions of digital tapes, or both. Documentation must include interpretation of at least the limiting cases.

Industrial use of transient studies has been limited primarily to arc furnace melt shop systems which entail a high rate of furnace switching, usually via vacuum switches, and usually in the presence of significant amounts of power capacitors which are sometimes also switched. In these applications study objectives typically determine:

(1) Transient and harmonic sensitivities associated with furnace and capacitor switching

(2) Preferred switching sequences

(3) If tuning reactors for capacitors are justifiable

(4) Effect of resistor insertion

(5) Proper surge protective equipment specifications

(6) Transient duties associated with inception of faults (line to ground and multi-phase) and fault clearing

10.9 Switching Transient Study Via Transient Network Analyzer (TNA). The TNA is the most used method of conducting calculated predictions of transient behavior in a power system. Accordingly the TNA will be elaborated upon in this section, followed by an example of a case selected from an actual industrial system study. The TNA uses electromagnetic models primarily in representing the system to be studied whereas digital and analog or hybrid computer analysis is based on mathematical modeling. The hybrid simulator, while composed of general purpose components, can still be special purpose to a degree depending on special components that are included. The TNA is basically special purpose in nature, being devoted virtually entirely to switching transient analysis. The following discusses salient aspects of a TNA facility and associated study procedures.

10.9.1 Model Components. Power for the model system is supplied from a low-impedance three-phase sine-wave source, sufficiently large to appear as a virtual infinite source to the model system. System equivalent impedances at points of infeed are represented by linear reactances supplied from the infinite source in the associated positive, negative, and zero sequence component networks. In a similar manner, generators are usually represented by their direct-axis transient or subtransient reactances. Transfer impedances between buses are represented when their contribution is considered significant.

Because of the effect of traveling waves, transmission lines and cables are represented by a quasi-distributed-constant model composed of a number of cascaded three-phase, four-wire L or pi sections. Each section consists of a series of resistance and inductance plus shunt capacitance to simulate positive and zero sequence line constants. Depending on system configuration and objectives of the study, each section can be made to represent widely varying lengths of circuit. Models representing completely untransposed lines or coupled lines on the same right-of-way, or both, can be modeled in greater detail when these effects are considered sufficient to have

a significant influence on the transients.

Two- or three-winding transformers or autotransformers are represented by models which accurately match positive, negative, and zero sequence leakage impedances; as well as complete positive and zero sequence saturation characteristics for single- or three-phase core or shell-type designs. Phase-shifting transformers can also be modeled.

Series reactors are modeled using model reactors which are linear. Shunt reactors are modeled as linear or saturable, as required.

Shunt capacitor banks can be modeled using miniature model capacitors.

Series capacitors can be accurately modeled, including their protective bypass and reinsertion switches.

Harmonic filters, including *tuned* capacitor banks for harmonic resonance control, are modeled using combinations of model resistors, reactors, and capacitors.

It is possible to accurately represent any surge arrester design, including the complex, current-limiting valve type and zinc oxide valve type. However, detailed application of each type is usually not warranted; instead, a relatively simple representation consisting of a model voltage-sensitive gap in series with a nonlinear resistor forms an adequate base from which evaluations of specific surge arrester designs can be made.

Circuit breakers or other switching devices are represented by either a mechanical or electronic-controlled multi-contact switch which operates in synchronism with the TNA power supply. This TNA switch can be adjusted to independently represent the three poles of the power system switching device for either opening or closing operations and can be direct acting or can accommodate single- or multi-step resistor pre-insertion.

Where loads significantly influence transients, a static representation via combinations of model resistors, reactors, and capacitors is made. Load flows in a multi-terminal system are represented through phase-angle regulators at appropriate model system infeeds.

For convenience of interconnection and instrumentation, all various individual components are connected to a plug board or patch panel.

10.9.2 Model Scale Factors. Generally, a voltage scale factor is introduced on transient network analyzer representations. For example, 100 V on the TNA model might represent 34.5 kV on the actual system.

Transient network analyzer representations of systems can be made with model elements of the same ohmic values as their system counterparts. However, to minimize the number of model elements required for system parameters, other electrical scale factors are introduced in accordance with established theories of modeling. Impedance scale factors are given for convenience. Time scale factors have been advantageous for some studies.

10.9.3 Accuracy. In general, where reliable measurements of transients in the field have been conducted under conditions and switching circumstances duplicating that of TNA analysis, correlation of results has been remarkably good. Of course, accuracy of TNA results depends on the accuracy of data and the degree to which models can be made to approach the desired characteristics. In the typical industrial system, modeling system components is usually a straightforward procedure with transformers being the most complex and requiring the most data. Modeling switch characteristics confined to a limited number of reconductions (reignitions, preignitions, restrikes) is

within the capability of the *conventional* TNA switching representation. However, modeling numerous reignitions in a switching device within a very small part of a cycle of fundamental frequency, such as may occur in some vacuum switching circumstances, requires electronic augmentation of the TNA switch representation.

High voltage transmission transient measurements have confirmed TNA predictions within 4% magnitude deviation. Limited transient measurements on arc furnace systems in industry indicate TNA predictions to be modestly higher, say within 10%, than actual prevailing transient duties. This, to a large extent, is expected because TNA results are most often maximized, that is, assuming the most unfavorable combination of switching device pole-closing electrical angles. As such they are expected to be higher than that indicated by field measurements where it is unlikely that the worst (most pessimistic) pole closing angle combination will occur, at least during a staged field test.

It is mandatory that all TNA model components possess the same high-frequency characteristic as individual items of system apparatus that they represent. Thus the models must entail appropriate capacitive, inductive, and resistive elements to duplicate the frequency response over the range of frequencies of interest, model electrical scale factors being considered. Perhaps the most difficult quantity to model is the resistance of some reactors and transformers, which tends to be proportionately higher in the model than in the represented equipment. This circumstance is a challenge to the TNA modeling specialist. In fact, the reasonably precise simulation of transformers is generally the most difficult area of TNA modeling. Nonlinear magnetic characteristics, losses, leakage inductance, distributed capacitance — all must be accommodated within reasonable limits. Electronic circuit modeling is sometimes used to achieve desired nonlinearities and losses and to exhibit correct terminal behavior.

10.9.4 Study Procedures. After all model components have been interconnected to properly represent the actual system to be studied, the switching operation is performed. With this model transient voltages and currents can be observed and measured on an oscilloscope or by means of digital metering by connecting measuring probes directly to various locations in the model system.

For energizing, reclosing, or restriking investigations, positions of the individual closing contacts on the TNA switch (representing the three phases of the power system switching device) are varied and adjusted both with respect to the driving voltage sine wave and to each other. For a given set of conditions there exists a particular combination of pole-closing angles which produces the maximum closing transient. Searching out this maximum transient pole-closing combination within the allowable time span between poles is called *maximizing* the transient. Maximizing is done manually and requires considerable practice to accomplish it efficiently and dependably. Maximizing accomplishes in a few minutes what otherwise could require many computer runs if done digitally.

For investigations of overvoltages following switching device opening operations, the model system is first energized, and after all energizing transients have subsided, contacts of the TNA switch are adjusted to interrupt on successive current zeros, just as on an actual circuit breaker.

Analog computer controlled switching is used to generate a statistical overvoltage distribution on the TNA. One such facility utilizes two analog computers, one to simulate natural variations in actual switching device closing angles and closing span, and the other to establish actual overvoltage distribution produced on the TNA. Experience with many distribution curves and corresponding hand-maximized transients indicates that a hand-maximixed transient is attained in only a small fraction of 1% of the switching events in actual system service.

Where surge arrester operation is expected to make important modifications of the transient under consideration, the TNA model surge arrester can be sparked over at some point between reseal rating and maximum guaranteed sparkover voltage of the arrester for maximum magnitude to successive voltage crests in the modified transient voltage wave. Arrester discharge currents and the modified transients are observed and recorded. Such records are examined to determine the minimum applicable surge arrester ratings consistent with reasonable system operating procedures.

A permanent record of the transients is usually made by photographing the traces on the oscilloscope screen.

10.9.5 Example of TNA Study Case Documentation. Figure 99 illustrates one particular form of a TNA study case data sheet. The left half illustrates an abbreviated one-line diagram containing essential components of the case. Provision for describing the switching event and the system arrangement along with the disposition of loads is included immediately below the one-line diagram. For the case illustrated, the switching event is the simultaneous energizing of a capa-

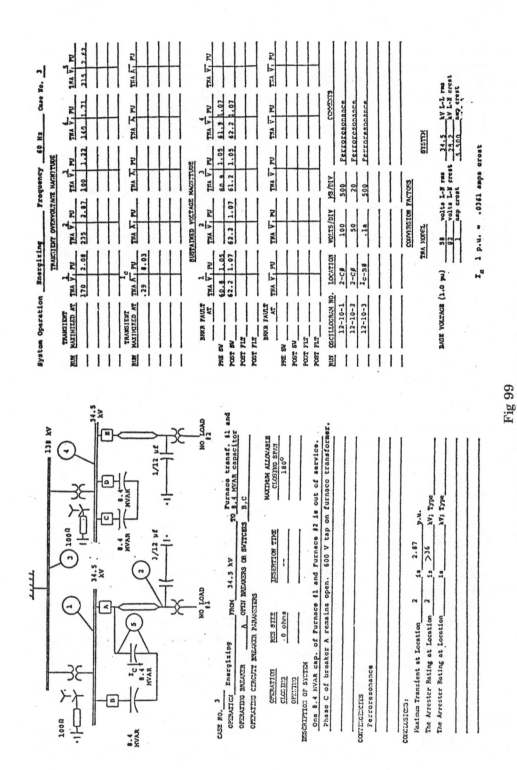

Fig 99
Example of TNA Case Sheet from an Actual
Switching Transient Study

citor and a furnace transformer by closing a breaker wherein one pole (phase C) remains open. The description also indicates the breaker used no resistor insertion and a pole-closing span of 180° was searched to attain a maximized transient of 2.87 per unit voltage which occurred at the energized transformer primary (that is, location 2; the measuring probe locations are indicated by circles).

The right half of the case sheet consists of a tabulated format so transient and sustained voltage magnitudes can be recorded up to ten probe locations each. Provision is made to show both pre-switching and post-switching (and pre-fault and post-fault) sustained voltage levels. Also, an index is included for oscillograms (which in the TNA report are generally reproduced on the page(s) immediately following the case sheet). Oscillograms are photos of CRO displays, generally a separate photo for each phase.

At the bottom right half sheet there is a listing of electrical scale factors used in the modeling, that is, the relationship between the TNA model and system electrical quantities.

Note in the case sheet illustrated in Fig 99, only one run was conducted and the most severe overvoltage was 2.87 per unit as a result of ferroresonance which developed as a consequence of breaker malfunction (one pole remaining open). The magnitude of this overvoltage is determined by scaling the oscillogram which displays the greatest instantaneous voltage or by direct read-out from a digital peak-holding voltmeter. Figure 100 shows an oscillogram which discloses the characteristic ferroresonant pattern as well as the magnitude of associated peaks. It also discloses that the predominant frequency of the ferroresonance is 60 Hz. Those familiar with switching transient analysis will immediately recognize that ferroresonance has developed as a result of a 60 Hz impedance match (or a near match) between the magnetizing reactance of the transformer being energized and its $1/12$ μF surge capacitor. This inductive-capacitive combination is excited by a zero-sequence driving voltage occasioned by the open breaker pole. The multi-frequency nature of the oscillogram is a result of the nonlinear magnetizing reactance of a transformer, particularly in the period immediately

Fig 100
Oscillogram 12–10–2 as Described
in TNA Case Sheet of Fig 99

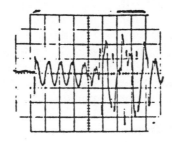

following energizing.

Complete switching transient study documentation includes not only detailed individual case study sheets for transient responses associated with various arrangements and conditions surveyed, but also analysis, recommendations, and conclusions of the study. Finally, the study documentation should include a complete listing of parameters (R, L, and C) of various system components, characteristics of protective devices, and a description of any unusual or special representations used in the study.

10.10 Field Measurements. The important role of field measurements in switching transient investigations has been noted earlier in this section. The choice of measurement equipment, auxiliary equipment selection, and techniques of setup and operation is in the domain of practiced measurements specialists. No attempt will be made here to delve into such matters in detail, except from the standpoint of conveying the depth of involvement entailed by switching transient measurements and from the standpoint of planning a measurements program to secure reliable transient information of sufficient scope for purposes intended.

10.10.1 Signal Derivation. Field measurements are conducted with staged tests wherein a prescribed sequence of switching operations are conducted for various system operating modes, or they may relate to monitoring day-by-day operating conditions. In any case, switching transients involve natural frequencies of orders of magnitude such that signal sourcing must be by special current transformers CTs, noninductive resistance dividers, non-inductive shunts, or compensated capacitor dividers. While con-

ventional CTs and potential transformers PTs can be suitable for harmonic investigations involving up to only a few kilohertz, their frequency response is usually inadequate for switching transient measurements.

In transient switching field measurements, voltage signals are derived from the power circuit via compensated capacitor divider and conveyed to the measuring equipment via shielded (or double shielded) coaxial cable. Sometimes so-called *captap* dividers are used for deriving voltage signals at apparatus bushings provided with capacitance taps or power factor taps. The captap device consists of a known value of capacitance made up of capacitors arranged in a coaxial array and fitted into a shielding container. This container includes a special adapter which is screwed into the capacitance tap or power factor tap of the respective bushing. This external known capacitance, together with the inherent known capacitance of the bushing, forms a capacitive voltage divider. The divider is formed by the internal line-to-tap bushing capacitance and the much greater capacitance of the external adapter capacitance from tap to ground. The geometry of the adapter can be made such that this bushing-adapter system is capable of responding to very short-time wavefronts, less than 100 nanoseconds (0.1 μs). This response is more than adequate for faithful recording of switching transient voltages encountered in power systems.

10.10.2 Signal Circuits, Terminations, and Grounding. Due to very high currents with associated high magnetic flux concentrations that may attend transient phenomena in power systems, it is essential that signal circuitry be extremely well shielded and constructed to be as interference-free as possible. Double-shielded coaxial low-loss signal cable is satisfactory for this purpose. Additionally, it is essential that signal circuit terminations be made carefully with high quality hardware and be carefully impedance matched to avoid spurious reflections. It is desirable that signal circuits and instruments be laboratory tested as an assembly before field measurements are undertaken. This testing should include the injection of a known steep wave into the input end of the signal circuit and comparison of this wave shape with that on the receiving instruments (scopes). Only after a close agreement between the two wave shapes is achieved should the assembly be approved for switching transient field measurements. These tests also aid overall calibration.

All the components of the measurements system should be grounded via a continuous conducting grounding system of lowest practical inductance to minimize internally induced voltages. The grounding system should be configured to avoid so-called ground loops which can result in noise injection. Where signal cables are unusually long, excessive voltages can become induced in their shields, but industrial switching transient measurement systems have not as yet involved such cases.

10.10.3 Transient Measurement/Monitoring Instrumentation. The complement of instruments used depends on circumstances and purpose of the test program. Major items comprising the total complement of display and recording instrumentation for transient measurements are one or more of the following:

(1) One or more oscilloscopes including a storage-type scope with multi-channel switching capability. When the presence of the highest speed transients (that is, with front times of a small fraction of a microsecond) is suspected, a high-speed, single-trace surge test oscilloscope with direct CRT (cathode ray tube) connections is sometimes used to record such transients with least possible distortion

(2) A multi-channel magnetic light beam oscillograph with high input impedance amplifiers

(3) A peak-holding digital-readout memory voltmeter (sometimes called *peak picker*), usually manually reset

The storage scope should have at least three-channel capability to permit simultaneous display of the three phase-voltage signals. An additional channel is desirable to allow a spare or to display another signal of interest. The single-trace surge test oscilloscope with direct CRT input is capable of producing the highest resolution of specific signals of interest on faster sweep speeds, normally from 10 to 200 μs/divisions.

From the standpoint of conducting switching transient field measurements, one of the most difficult aspects is securing an acceptable and reliable triggering method for the storage scope when multichannel switching is used to record more than one signal. Considerable experimenting may be necessary in order to *catch* the transient activity due to its short duration. One successful approach in some tests on systems with open (nonshielded) bus has been to use a simple wire antenna connected to the external trigger of the scope. The antenna will sense air-born signals emanating from the power circuit bus in concert with initiation of the switching. Associated sweep speeds of 200 to 1000 μs/divisions have been found generally most useful for recording all but the very fastest switching transient voltages.

The magnetic oscillograph displays all voltages and signals being monitored. Current signals derived from special current transformers or shunts are fed directly to oscillograph galvanometers through appropriate damping resistors. Voltage signals derived from capacitive dividers are isolated from low-impedance galvanometers by high input-impedance (megohm range) oscillograph amplifiers. Oscillograph records are virtually indispensable in the efficient interpretation of transient phenomena recorded through the orientation of the broad perspective they provide. Such oscillograph traces also give records of slow-speed switching transients, system oscillations, and harmonics.

Finally, the peak-holding voltmeter allows a valuable quantitative on-the-spot evaluation of the severity of transients produced by a particular switching operation. This permits a quick comparison between test runs and also can be left on for extended periods, sometimes unattended, to obtain a reading of the highest transient occurring during that period.

10.10.4 Objectives of Field Measurements. See the foregoing section on switching transient study objectives. Field measurements seldom, if ever, include fault switching, and often recommended corrective measures are not in place to be put into the test program, except on a follow-up basis.

11. Reliability Studies

11.1 Introduction. An important aspect of power system design involves consideration of service reliability requirements of loads to be supplied and service reliability provided by any proposed system. System reliability assessment and evaluation methods based on probability theory allow the reliability of a proposed system to be assessed quantitatively. Such methods permit consistent, defensible, and unbiased assessments of system reliability which are not otherwise defensible, and which are not otherwise possible.

Quantitative reliability evaluation methods permit reliability indexes for any electric power system computed from knowledge of the reliability performance of the constituent components of the system. Thus, alternative system designs can be studied to evaluate the impact on service reliability and cost of changes in component reliability, system configuration, protection and switching scheme, or system operating policy including maintenance practice. A detailed treatment of reliability evaluation methods is given in the IEEE Recommended Practice for Reliability Evaluation of Industrial and Commercial Power Systems [1].

11.2 Definitions. The definitions presented here provide much of the required nomenclature for discussions of power system reliability.

availability. A term which applies either to the performance of individual components or to a system. Availability is the long-term average fraction of time that a component or system is in service satisfactorily performing its intended function. An alternative and equivalent definition for availability is the steady-state probability that a component or system is in service.

component. A piece of equipment, a line or circuit, or a section of a line or circuit,

191

or a group of items which is viewed as an entity for purposes of reliability evaluation.

expected interruption duration. The expected, or average, duration of a single load interruption event.

exposure time. The time during which a component is performing its intended function and is subject to failure.

failure. Any trouble with a power system component that causes any of the following to occur:

(1) Partial or complete plant shutdown, or below-standard plant operation

(2) Unacceptable performance of user's equipment

(3) Operation of the electrical protective relaying or emergency operation of the plant electrical system

(4) Deenergization of any electric circuit or equipment

A failure on a public utility supply system can cause the user to have either of the following:

(1) A power interruption or loss of service

(2) A deviation from normal voltage or frequency of sufficient magnitude or duration

A failure on an in-plant component causes a forced outage of the component, that is, the component is unable to perform its intended function until repaired or replaced. The terms *failure* and *forced outage* are often used synonymously.

failure rate (forced outage rate). The mean number of failures per unit of exposure time for a component. Usually *exposure time* is expressed in years and *failure rate* is given in terms of failures per year.

forced unavailability. The long-term average fraction of time that a component or system is out of service as a result of failures.

interruption. The loss of electric power supply to one or more loads.

interruption frequency. The expected average number of power interruptions to a load per unit time, usually expressed as interruptions per year.

outage. The state of a component or system when it is not available to properly perform its intended function.

repair time. The clock time from the time of component failure to the time when the component is restored to service, either by repair of the failed component or by substitution of a spare component for the failed component. It is not the time required to restore service to a load by putting alternate circuits into operation. It includes time for diagnosing the trouble, locating the failed component, waiting for parts, repairing or replacing, testing, and restoring the component to service. The terms *repair time* and *forced outage duration* can be used synonymously.

scheduled outage. An outage that results when a component is deliberately taken out of service at a selected time, usually for purposes of construction, maintenance, or repair.

scheduled outage duration. The time period from the initiation of a scheduled outage until construction, preventive maintenance, or repair work is completed and the affected component is made available to perform its intended function.

scheduled outage rate. The mean number of scheduled outages per unit of exposure time for a component.

switching time. The period from the time a switching operation is required because of a component failure until that switching operation is completed. Switching operations include such operations as: throwover to an alternate circuit, opening or closing a sectionalizing switch or circuit breaker, reclosing a circuit breaker following a trip-out from a temporary fault, etc.

system. A group of components connected or associated in a fixed configuration to perform a specified function of distributing power.

unavailability. The long-term average fraction of time that a component or system is out of service caused by failures or scheduled outages. An alternative definition is the steady-state probability that a component or system is out of service. Mathematically, unavailability = (1 – availability).

11.3 System Reliability Indexes. The two basic system reliability indexes which have proven most useful and meaningful in power distribution system design are load interruption frequency and expected duration of load interruption events. These indexes can be readily computed using the methods in [1]. The two basic indexes of interruption frequency and expected interruption duration can be used to compute other indexes which are also useful:

(1) Total expected average interruption time per year, or other time period

(2) System availability or unavailability as measured at the load supply point in question

(3) Expected energy demanded, but unsupplied, per year

Note that the disruptive effect of power interruptions is often non-linearly related to the duration of the interruption. Thus, it is often desirable to compute not only an overall interruption frequency but also frequencies of interruptions categorized by the appropriate durations.

11.4 Data Needed for System Reliability Evaluations. Data needed for quantitative evaluation of system reliability depends to some extent on the nature of the system being studied and the detail of the study. In general, however, data on the performance of individual components together with the times required to perform various switching operations are required.

System component data generally required are summarized as follows:

(1) Failure rates (forced outage rates) associated with different modes of component failure

(2) Expected average time to repair or replace failed component

(3) Scheduled maintenance outage rate of component

(4) Expected average duration of a scheduled outage event

If possible, component data should be based on historical performance of components in the same environment as those in the proposed system being studied. The reliability surveys conducted by the Power Systems Reliability Subcommittee [2], [3] allow a source of component data when such specific data is not available.

Switching time data needed includes:

(1) Expected times to open and close a circuit breaker

(2) Expected times to open and close a disconnect or throwover switch

(3) Expected time to replace a fuse link

(4) Expected times to perform such emergency operations as cutting in clear, installing jumpers, etc

Switching times should be estimated for the system being studied based on experience, engineering judgment, and anticipated operating practice:

11.5 Method for System Reliability Evaluation. The general method for system reliability evaluation which is recommended has evolved over a number of years. The method is well suited to the study and analysis of electric power distribution systems as found in industrial plants and commercial buildings. The method is systematic and straightforward and lends itself to either manual or computer computation. An important feature of the method is that system weak points can be readily identified, both numerically and non-numerically, thereby focusing design attention on those sections of the system which contribute most to service unreliability.

The procedure for system reliability evaluation is outlined as follows:

(1) Assess the service reliability requirements of the loads and processes supplied and determine appropriate service interruption definition or definitions

(2) Perform a failure modes and effects analysis (FMEA) identifying and listing those component failures and combinations of component failures which result in service interruptions and constitute minimal cut-sets of the system

(3) Compute interruption frequency contribution, expected interruption duration, and the probability of each of the minimal cut-sets of (2)

(4) Combine results of (3) to produce system reliability indexes

The above steps are discussed in more detail in later sections.

11.5.1 Service Interruption Definition. The first step in any electric power system reliability study should be a careful assessment of the power supply quality and continuity required by the loads which are served. This assessment should be summarized and expressed in a service interruption definition used in the succeeding steps of the reliability evaluation procedure. The interruption definition specifies the reduced voltage level

(voltage dip) together with the minimum duration of such reduced voltage period which results in substantial degradation or complete loss of function of the load or process being served. Frequency reliability studies are conducted on a *continuity* basis in which case interruption definitions reduce to a minimum duration specification with voltage assumed to be zero during the interruption.

11.5.2 Failure Modes and Effects Analysis. Failure modes and effects analysis (FMEA) for power distribution systems amount to determination and listing of those component outage events or combinations of component outages which result in an interruption of service at the load point being studied according to the interruption definition adopted. This analysis must be made considering the different types and models of outages which components can exhibit and the reaction of the system's protection scheme to these events. Component outages are categorized as:

(1) Forced outages or failures

(2) Scheduled or maintenance outages

(3) Overload outages

Forced outages or failures are either permanent forced outages or transient forced outages. Permanent forced outages require repair or replacement of the failed component before it can be restored to service while transient forced outages imply no permanent damage to the component thus permitting its restoration to service by a simple re-closing or re-fusing operation. Additionally, component failures can be categorized by physical mode or type of failure. This type of failure categorization is important for circuit breakers and other switching devices where the following failure modes are possible:

(1) Faulted, must be cleared by back-up devices

Table 16
Frequency and Expected Duration Expressions for
Interruptions Associated with Forced Outages Only

First-Order Minimal Cut-Set	Second-Order Minimal Cut-Set
$f_{cs} = \lambda_i$	$f_{cs} = \lambda_i \lambda_j (r_i + r_j)$
$r_{cs} = r_i$	$r_{cs} = r_i r_j / (r_i + r_j)$

Third-Order Minimal Cut-Set

$$f_{cs} = \lambda_i \lambda_j \lambda_k (r_i r_j + r_i r_k + r_j r_k)$$
$$r_{cs} = r_i r_j r_k / (r_i r_j + r_i r_k + r_j r_k)$$

Symbols: f_{cs} = frequency of cut-set event

r_{cs} = expected duration of cut-set event

λ_i = forced outage rate of ith component

r_i = expected repair or replacement time of ith component

NOTE: The time units of r and λ in expressions for f_{cs} must be the same. Once frequencies and expected durations have been computed for each minimal cut-set, system reliability indexes at the load point in question are given by:

f_s = interruption frequency

$$= \sum_{\substack{\min \\ \text{cut-sets}}} f_{cs_i}$$

r_s = expected interruption duration

$$= \sum_{\substack{\min \\ \text{cut-sets}}} f_{cs_i} r_{cs_i} / f_s$$

$f_s r_s$ = total interruption time per time period

(2) Fails to trip when required

(3) Trips falsely

(4) Fails to re-close when required

Each will produce a varying impact on system performance.

The primary result of the FMEA as far as quantitative reliability evaluation is concerned is the list of minimal cut-sets it produces. A minimal cut-set is defined to be a set of components which if removed from the system results in loss of continuity to the load point being investigated and does not contain as a subset any set of components that is itself a cut-set of the system. In the present context the components in a cut-set are just those components whose overlapping outage results in an interruption according to the interruption definition adopted.

An important non-quantitative benefit of FMEA is the thorough and systematic thought process and investigation it requires. Often weak points in system design are identified before any quantitative reliability indexes are computed. Thus, the FMEA is a useful reliability design tool even in the absence of the data needed for quantitative evaluation.

11.5.3 Computation of Quantitative Reliability Indexes. Computation of reliability indexes can proceed once the minimal cut-sets of the system have been found. The first step is to compute the frequency, expected duration, and expected down-time per year of each minimal cut-set. Note that expected down-time per year is the product of the frequency expressed in terms of events per year and the expected duration. If the expected duration is expressed in years, the expected down-time will have the units of years per year and can be regarded as the relative proportion of time or probability the system is down due to the minimal cut-set in question. More commonly, expected duration is expressed in hours and the expected down-time has the number of hours per year.

Approximate expressions for frequency and expected duration of the most commonly considered interruption events associated with first-, second-, and third-order minimal cut-sets are given in Table 16. Note that expressions for the calculaforced outages (failures) only. A detailed treatment of expressions for the calculation of interruption frequency and duration considering forced outages as well as

maintenance outages, switching after faults to restore service, and incomplete redundancy of parallel facilities is given in [1].

11.6 References

[1] IEEE Std 493-1980, Recommended Practice for the Design of Reliable Industrial and Commercial Power Systems.

[2] IEEE COMMITTEE REPORT, Report on Reliability Survey of Industrial Plants, Part I: Reliability of Electrical Equipment, *IEEE Transactions on Industry Applications*, pp. 213–235, March/April 1974.

[3] IEEE COMMITTEE REPORT, Reliability of Electric Utility Supplies to Industrial Plants, Conference Record 1975 I & CPS Technical Conference, pp. 131–133.

12. Grounding Mat Studies

12.1 Introduction. A grounding mat study has one primary purpose: to ensure that the ground mat design provides for the safety and well being of linemen, maintenance personnel, operators, and the public, anyone who comes near an electrically conductive object tied to a ground mat under the influence of a major ground fault. Equipment protection or system operation is rarely an objective of a grounding mat study, which makes this particular type of analysis rather unique. An increasing concern of industry and utilities for the safety of their employees has led to greater interest in predicting voltage characteristics of grounding mats experiencing heavy fault currents.

12.2 The Human Factor. To properly understand the analytical techniques involved in a grounding mat study, it is necessary to understand the electrical characteristics of the most important part of the circuit — man. A normally healthy person can feel a current of about 1 mA.[3] Currents of approximately 10–25 mA can cause lack of muscular control. In most men 100 mA will cause ventricular fibrillation. Higher currents can stop the heart completely or cause severe electrical burns.

For practical considerations, the threshold of fibrillation is the major concern of most grounding mat studies. Ventricular fibrillation is a condition where the heart beats in an abnormal and ineffective manner, with fatal results. Accordingly, most ground mats are designed to limit body currents to values below this threshold. Probably the most useful information on this subject can be found in [1],[1] [2], [3], and [4] (the figures cited in the preceding paragraph were taken

---

[1]Numbers in brackets correspond to those in the References at the end of this Section.

[3]Tests have long ago established the now well-known fact that electric shock effects are the result of current and not voltage [5].

from these references). Although the test results on ventricular fibrillation were actually taken from animals with body and heart weights comparable to those of a man, the results have been generally accepted as being valid for human beings. These studies have determined that 99.5 percent of all healthy men can tolerate a current through the heart region defined by

$$I_b = \frac{0.116}{\sqrt{T}} \qquad \text{(Eq 1)}$$

where

I_b = maximum body current in amperes

T = duration of current in seconds

without going into ventricular fibrillation. Obviously, this equation precludes choosing a single value for the fibrillation threshold current. Even a high current through the heart region can be tolerated for a brief period. Other tests show that this threshold current is approximately five times greater for direct current [5] and as much as 25 times greater for 3000 Hz [6]. Therefore, Eq 1 should embody sufficient conservatism for all cases where an individual might be exposed to 60 Hz ac fault potentials (the dc component of asymmetrical fault current is accurately recognized by a suitable *correction factor* described in a later section).

Tests indicate that the heart requires about five minutes to return to normal after experiencing a severe shock [1]. This implies that two or more closely spaced shocks (such as those that would occur in systems with automatic reclosing) would tend to have a cumulative effect. Present industry practice considers two closely spaced shocks to be equivalent to a single shock whose duration is the sum of the intervals of the individual shocks. The same series of tests also showed that the body can tolerate much more current flowing from one leg to the other than it can when current flows from one hand to the legs.

Figures 101 and 102 show two typical shock hazard situations and the equivalent resistance diagrams. Figure 101 shows a touch contact with current flowing from operator's hand to his feet. Figure 102 shows a step contact where current flows from one foot to the other. In each case the body current I_b is driven by the potential difference between points A and B. Exposure to touch potential normally poses a greater danger than exposure to step potential. The step potentials are usually smaller in magnitude, the corresponding body resistance greater, and the permissible body current higher than for touch contacts. (The current magnitude in the heart region that causes fibrillation is the same for both types of contacts. In the case of step potentials, however, not all current flowing from one leg to the other will pass through the heart region.) The worst possible touch potential (called mesh potential) occurs at or near the center of a grid mesh. Accordingly, industry practice has made mesh potential the standard criterion for determining safe ground mat design. In most cases, controlling mesh potential will bring step potentials well within safe limits inside the area defined by the grounding mat. Step potentials can, however, reach dangerous levels at points immediately outside the grid.

Since the body of a man exposed to an electrical shock forms a shunt branch in an electrical circuit, the resistance of this branch must be determined to calculate the corresponding body current. Generally, the hand and foot contact resistances are considered to be negligible.

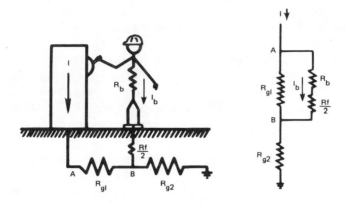

Fig 101
Touch Potential

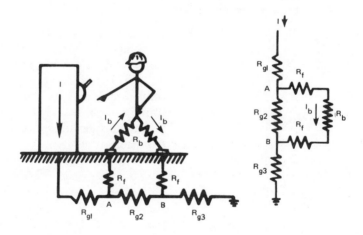

Fig 102
Step Potential

However, resistance of the soil directly underneath the foot contact area is considered significant. Treating the foot as a circular plate electrode gives an approximate resistance of $3\rho_s$ where ρ_s is the soil resistivity [3]. The body itself has a total measured resistance of about 2300 Ω hand to hand or 1100 Ω hand to foot.

In the interest of simplicity and conservatism, IEEE Std 80-1976 [6] recommends the use of 1000 Ω as a reasonable approximation for body resistance. This yields a total branch resistance of

$$R = 1000 \ \Omega + 6\rho_s \qquad \text{(Eq 2)}$$

for foot to foot currents and

$$R = 1000 \ \Omega + 1.5\rho_s \qquad \text{(Eq 3)}$$

for hand to foot currents where ρ_s is the surface resistivity in ohm meters and R is expressed in ohms. If the station surface has been dressed with crushed rock or some other high resistivity material, resistivity of the surface layer material should be used in Eq 2 and Eq 3.

Because potential is easier to both calculate and measure than current, the fibrillation threshold given by Eq 1 is normally expressed in terms of voltage. Combining Eq 1, Eq 2, and Eq 3 gives the maximum *tolerable* step and touch potentials:

$$E_{\text{step}} = \frac{(1000 \ \Omega + 6\rho_s)(0.116)}{\sqrt{T}} \qquad \text{(Eq 4)}$$

$$E_{\text{touch}} = \frac{(1000 \ \Omega + 1.5\rho_s)(0.116)}{\sqrt{T}} \qquad \text{(Eq 5)}$$

12.3 The Physical Circuit. Although in each of the cases discussed above man's body resistance shunts a part of the ground resistance, its actual effect on voltage and current distribution in the overall system is negligible. This becomes obvious when the normal magnitude of the ground fault current (as much as several thousand amperes) is compared to the desired body current (usually no more than several hundred milliamperes). Voltage rise of any point within the grid

therefore depends solely upon the characteristics of the permanent physical installation. Grid voltages depend upon three basic factors: ground resistivity, available fault current, and grid geometry. Proper consideration of each of these factors allows the analyst to recognize both hazardous and overly conservative designs.

12.3.1 Ground Resistivity. Obviously, the simplest grounding mat analysis involves a grid in a homogeneous medium. Unfortunately, substations must be located according to factors other than the ease of calculating grid voltages. Fortunately for those concerned with such calculations, the homogeneous medium assumption is sufficiently accurate for most soils. Also, a number of nonhomogeneous soils can be modeled by two-layer techniques [7], [8]. Although reasonably straightforward, these methods involve many tedious calculations making computation by hand difficult. The two-layer model is necessary only for locations where bedrock and other natural soil layers are close enough to the surface to severely affect the distribution of current.

Of far more serious concern are soils which experience drastic and unpredictable changes in resistivity at the earth's surface. These situations present problems:

(1) Difficulty of modeling soil in calculations

(2) Physical difficulties in finding discontinuities in the field and measuring local soil resistivity.

At present, these cases are normally handled by the inclusion of a safety margin in the value used for the soil resistivity.

A description of a simple method of measuring soil resistivity of homogeneous soils is given in [6]. Techniques also

Table 17
Representative Values of Soil Resistivities

Type of Ground	Resistivity in ohm-meters
Wet organic soil	10
Moist soil	10^2
Dry soil	10^3
Bed rock	10^4

exist for measuring the resistivity of each layer of two-layer soils [8]. Because soil resistivity varies with moisture content and, to a lesser degree, with temperature, these measurements should be made over a period of time under different weather conditions. If for some reason actual measurement of resistivity is impractical, tables of approximate values of resistivity for soils of various composition have been compiled by several sources. Table 17 is a sample table [9]. These values are only approximations and should be replaced in the study by more accurate figures whenever possible.

12.3.2 Fault Current—Magnitude and Duration. Determination of ground fault current normally entails a separate study. Techniques and problems of making fault studies are covered in numerous sources. This section will only cover aspects peculiar to grounding grid studies.

After the system impedance and grid resistance have been determined, the maximum ground fault current (assuming a bolted fault) is given as:

$$I = \frac{3V}{3R_g + (R_1 + R_2 + R_0) + j(X_1'' + X_2 + X_0)}$$

(Eq 6)

where

I = maximum fault current, amperes (note that this is not the same as the current, I_b, in Eq 1)

V = phase to neutral voltage, volts

R_g = grid resistance to earth, ohms

R_1 = positive sequence system resistance, ohms

R_2 = negative sequence system resistance, ohms

R_0 = zero sequence fault path resistance, ohms

X_1'' = positive sequence subtransient system reactance, ohms

X_2 = negative sequence system reactance, ohms

X_0 = zero sequence fault path reactance, ohms

This current will, in general, be a sinusoidal wave with a dc offset. Since dc current can also cause fibrillation, the current value I must be multiplied by an

Table 18
Decrement Factor for Use in Calculating
Electrical Shock Effect of
Asymmetrical ac Currents

Shock and Fault Duration		Decrement Factor
Seconds	Cycles (60 Hz)	
0.008	½	1.65
0.1	6	1.25
0.25	15	1.10
0.5 or more	30 or more	1.0

appropriate *correction factor* for this effect. This multiplier is called the *decrement factor* in [6], and a table of approximate values for it is provided (reproduced in Table 18). For more accurate results, the exact value for the decrement factor D is given by the equation

$$D = \sqrt{\frac{1}{T}\left[T + \frac{1}{\omega}\cdot\frac{X}{R}\left(1 - e^{-\frac{2\omega T}{X/R}}\right)\right]}$$

where (Eq 7)

T = duration of fault, seconds

ω = system frequency in radians per second

X = total system reactance, ohms

R = total system resistance, ohms

The current value calculated in Eq 6 must be multiplied by this factor to find the effective fault current. Note that time T in Eq 7 is the same as that used in Eqs 1, 4, and 5. To determine the fault duration it is necessary to analyze the relaying scheme to find the interrupting time for the current calculated by Eq 6. Substitution of this time into Eqs 4 and 5 will fix the maximum allowable step and touch potentials at the appropriate values. These maximum allowable potentials are used to check the voltages actually present within the grid. If any voltages exceed maximum limits, the grid should be redesigned.

The choice of the clearing time of either the primary protective devices or the backup protection for the fault duration depends on the individual system. Designers must choose between the two on the basis of the estimated reliability of the primary protection and the desired safety margin. Choice of backup device clearing time is more conservative, but it will result in a more costly ground mat installation.

Study of maximum ground fault current alone is not sufficient. Any low magnitude ground fault current that might persist for several minutes or more also presents the very real danger of asphyxiation. A fault current capable of inducing body currents of 10 to 25 mA can cause muscular paralysis in a man, including his lungs. Since the majority of people resume normal respiration upon removal of the current, interruption times of a minute or so at this level should prevent any lasting injury from this particular effect. A grounding mat design can be checked for this hazard by first finding the ground fault current that will result in a body current of 10 mA, then considering the protective relaying scheme to determine the time required to detect and interrupt this current. If the time is one minute or less, asphyxiation should not be a danger.

An accurate estimation of the ultimate system fault capability is necessary to ensure a safe design throughout the life of the substation or plant. Normal industry experience indicates that fault currents rise as power systems are expanded and modernized. After the initial construction phase, however, changes to the ground mat are prohibitively expensive. Therefore, it is vital to consider the total future expansion in the initial ground mat design.

12.3.3 Fault Current—The Role of Grid Resistance. Accurate calculation of ground fault currents presupposes that an accurate and dependable value for the grid resistance can be calculated. More literature exists on this aspect of grounding mat design than any other, so finding an approximate value for the required grid resistance is not particularly difficult. Most of the equations and techniques developed to calculate grid resistance are based on several simplifying assumptions, however, and produce results that are, by necessity, both conservative and somewhat inaccurate. For the most part the accuracy of the grid resistance equations depends upon how well they account for the different grid configurations likely to be encountered.

A formula for a quick simple calculation of resistance when a minimum of design work has been completed is given in [6]

$$R = \frac{\rho}{4r} + \frac{\rho}{L} \qquad \text{(Eq 8)}$$

where

R = grid resistance to ground, ohms
ρ = soil resistivity, ohm meters
L = total length of grid conductors, meters
r = radius of a circle with area equal to that of the grid, meters

Grid resistance depends on soil resistivity, grid area, and total length of the conductors forming the grid. These variables influence the resistance so heavily, that Eq 8 does not consider any others. By inspecting Eq 8 it also becomes evident that adding grid conductors to a mat to reduce its resistance eventually becomes ineffective. As the conductors are crowded together, their mutual interference increases to the point where new conductors tend only to redistribute fault current around the grid, rather than lower its resistance.

The first term of Eq 8 gives the resistance of a circular plate with the same area as the grid. The second term allows for the grid's departure from the idealized plate model. The more the length of the grid conductors increases, the smaller this term becomes. This equation is ideal for the initial stages of a study where only the most basic data about the ground mat are available.

Another method for determining grid resistance (with greater accuracy than the previously described method) offers the advantage of lending itself to use in a computer program that can precisely calculate voltages at any location within a general grid configuration. It simply expresses grid-to-ground resistance as the total voltage rise of the grid (relative to a "remote" ground reference) divided by total fault current. This method can be applied to any grid configuration with any number of conductor elements. However, because an accurate determination of the total voltage rise depends on the solution to a potentially overwhelming array of involved mathematical expressions, this method is not well suited for hand calculations.

Since grid resistance is viewed as a measure of the grid's ability to disperse ground fault current, many designers are tempted to use resistance as an indicator of relative safety of a ground mesh. In general, however, there is no direct correlation between grid resistance and safety. At high fault currents, dangerous potentials exist within low resistance grids. The only occasion when a low grid resistance can guarantee safety is when maximum potential rise of the entire grid (that is, grid potential) is less than allowable touch potential.

12.3.4 Grid Geometry. The potential rise of points protected by a grounding mat depends on such factors as: grid burial depth, length and diameter of conductors, spacing between each conductor, distribution of current throughout the grid, proximity of the fault electrode and the system grounding electrodes to the grid conductors, along with many other considerations of lesser importance. A perfectly rigorous analysis of all these variables for a grounding grid with any number of conductor elements would, at the very least, involve solving a like number of (1) differential equations (to find the current distribution along each line) [8], [5], and (2) simultaneous linear algebraic expressions (to find the current distribution throughout the entire grid) [16]. Although unfortunate, it is hardly surprising that the quantitative effect of these factors upon touch and step potentials is one of the most infrequently discussed aspects of grid analysis.

Paradoxically, these factors include most of the elements of grid design that normally must be changed to control grid voltages. Obviously, any analytical method that cannot predict the effect of important changes in grid geometry is of limited utility for design purposes. Fortunately, some of these factors can be (and, for the sake of practicality, should be) safely ignored, while (unfortunately) others are vitally important to a reliable and accurate analysis.

Until recently, [6] provided the only practical formulas for computing the effects of the grid geometry upon the step and touch potentials (Eq 9 and Eq 10)

$$E_{\text{mesh}} = K_m K_i \rho \cdot \frac{I}{L} \qquad \text{(Eq 9)}$$

$$E_{\text{step}} = K_s K_i \rho \cdot \frac{I}{L} \qquad \text{(Eq 10)}$$

where

ρ = soil resistivity, ohm meters

I = maximum total fault current, amperes (adjusted for the decrement factor)

L = total length of grid conductors, meters

K_m = mesh coefficient

K_s = step coefficient

K_i = irregularity factor

Coefficients K_m and K_s are calculated by two reasonably simple equations based upon the number of grid elements, their spacing and diameter, and depth of burial of the grid. Many assumptions made in developing these equations were not meant to describe rigorously either industry practices or physical law, but were instead intended to simplify and make manageable what would otherwise be a very involved analytical procedure. Equations 9 and 10 incorporate an *irregularity factor* K_i to compensate for the inaccuracies introduced by these simplifying assumptions. Except for applications involving very simple grid configurations, proper selection of a value for K_i is totally dependent upon the experience and judgment of the designer. Most choose to err on the side of safety and make K_i large, which results in an over designed ground mat installation—usually safe, but often expensive.

Equations 9 and 10 yield a *single* value of E_{mesh} and E_{step} respectively, for any particular ground grid system. Values obtained for E_{mesh} and E_{step} are intended to represent the *worst case* condition for the ground grid system without providing the analyst with any information as to where (or how often) these worst case conditions exist within the system. After E_{mesh} and E_{step} of the grid are calculated, they are compared to the calculated values of the tolerable E_{touch} and E_{step} as determined from Eq 4 and Eq 5, respectively, in order to establish whether or not the design can be judged to be safe. If, in fact, E_{mesh} exceeds $E_{\text{touch-tolerable}}$ and/or E_{step} exceeds $E_{\text{step-tolerable}}$ it is sometimes possible by inspection of the grid to determine mesh locations where additional cross-conductors should be added in order to achieve a safe design. The more general approach, however, especially when either of the tolerable values is more than only slightly exceeded, is to uniformly increase the number of subdivisions in the original system of grid meshes so as to result in a lower calculated value of either E_{mesh} or E_{step} (or both). Accordingly, substantial overdesign and unnecessary investment in buried conductor may result.

Although this traditional method for calculating step and mesh potential has given long and faithful service, grounding mat analysis computer programs provide more information about the effectiveness of the ground mat design. Greater accuracy results because calculation by computer program does not require as many simplifying assumptions.

More specifically, the key to an accurate ground grid analysis is the consideration of each individual grid element, rather than the en masse treatment used in [6]. Fig 103, for example, shows a single grid element located some depth h

below the earth's surface. The element runs from point (x_1, y_1, z_1) to (x_1, y_2, z_1) and is radiating current to the surrounding earth at the linear current density σ_l (the current per unit length). By integrating σ_l over the length of the grid element, the current flux δ can be found at any desired point a as follows:

$$\delta = \int_{y_1}^{y_2} \frac{\sigma_l}{4\pi} \frac{dy}{R^2} r \qquad \text{(Eq 11)}$$

where

δ = current/unit area at any point
σ_l = current flowing to ground/unit length of conductor (current density)

$$R = \sqrt{(i - x_1)^2 + (j - y)^2 + (k - z_1)^2}$$

$$r = \frac{(i - x_1)i + (j - y)j + (k - z_1)k}{\sqrt{(i - x_1)^2 + (j - y)^2 + (k - z_1)^2}}$$

Fig 103
Physical Model Used in Calculating Voltage
at Point a Due to a Single Conductor

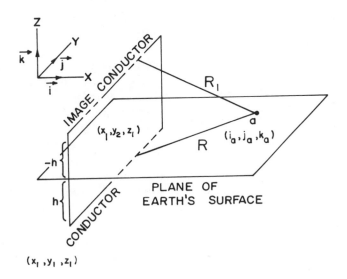

Once δ has been determined, the E-field at the same point a can be expressed as follows for a homogeneous soil:

$$E = \rho\delta \qquad \text{(Eq 12)}$$

where ρ is soil resistivity. From this, the voltage at point a can be obtained by performing the classical integration:

$$V_{a_l} = -\int_{\infty}^{a} E \cdot dl \qquad \text{(Eq 13)}$$

or,

$$V_{a_l} = -\frac{\rho\sigma_l}{4\pi} \ln\,(j_a - y_2$$

$$+ \sqrt{(i_a - x_1)^2 + (j_a - y_2)^2 + (k_a - z_1)^2})$$

$$+ \frac{\rho\sigma_l}{4\pi} \ln\,(j_a - y_1$$

$$+ \sqrt{(i_a - x_1)^2 + (j_a - y_1)^2 + (k_a - z_1)^2})$$

$$\text{(Eq 14)}$$

where V_{a_l} is the absolute potential at any point a due to line l. This process must then be repeated for every element in the grid.[4] Finally, the individual contribution of each grid element to the potential at point a must be summed to determine the total voltage at every such point of interest.

The advantages of this analytical method are immediately apparent. This technique automatically accounts for the finite length of each element, a particu-

larly important consideration when finding the potential at points near the end of an element (that is, at the corners of a ground mat). It can handle grid designs with large degrees of asymmetry with no sacrifice of accuracy (symmetry is not presupposed as with the calculating procedures in [6]). Further, since point a can be located anywhere and examined as often as required, detailed analysis of grid design is possible. The most significant disadvantage of this and other similar techniques is the number of tedious calculations that must be performed to accurately model a system. However, modern digital computers have all but eliminated this concern.

12.4 The Computer in Action. The utility and accuracy of the new computer programs can best be appreciated by examination of some actual test cases. Figure 104 summarizes the result of a computer analysis performed on six different grid layouts. Measurements recorded during tests on small scale physical models of these designs provide a convenient set of *benchmark* solutions that can be compared to the calculated results of a grounding mat analysis computer program.

The calculated voltages (expressed as a percentage of the grid voltage) are shown in the center of each mesh in Fig 104. (For easy reference, the actual measured voltages are also shown in parentheses.) Each value is the absolute voltage at the point shown relative to *remote* ground. To determine the touch potential hazard, the calculated point voltages must be subtracted from the calculated grid potential. For purposes of design or evaluation the resulting touch potential is directly compared to the maximum tolerable value. In the case of step potentials, the potential difference between any

---

[4] This process is complicated somewhat by the presence of the "current density" factor σ_l in the equations. Although Eq 11 and Eq 12 treat σ_l as a constant, in actuality it varies continuously along the length of each grid element. In practice, however, very accurate results can be obtained by using, on a piecewise basis, a constant value for σ_l that can be determined by one of several approximating techniques.

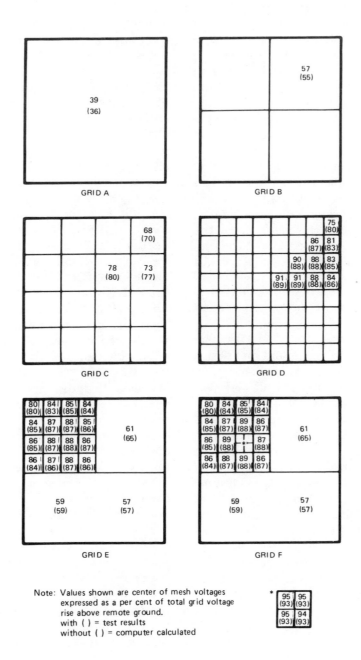

Note: Values shown are center of mesh voltages
expressed as a per cent of total grid voltage
rise above remote ground.
with ( ) = test results
without ( ) = computer calculated

**Fig 104
Experimental Grids Showing Various (Mesh) Arrangements**

two calculated point voltages (taken one meter apart) should be compared with the maximum allowable step voltage.

Figure 104 (especially grids E and F) illustrates the accuracy of the program. Errors are typically on the order of 5 percent, and never worse than 10 percent. Even the results for grids E and F, both of which are highly irregular, are well within acceptable design accuracy.

The most impressive feature of this program, however, is its ability to calculate voltages at any point of interest within or around the mat's geometric boundaries. By repeated use of the program throughout the design process, a grounding mat layout can be *fine-tuned* to achieve the desired protection without the need to overdesign any section of the mat.

This process is illustrated in Figs 105–108 using the typical ground mat design from [Appendix B, 18]. Figure 105 shows the grid layout and all pertinent grid information. Figure 106 gives the results of a computer analysis of the grid clearly defining the areas of maximum danger. Figures 107 and 108 show the results of modified grid designs. Note in Fig 108 that the amount of additional conductor required to safely control mesh potentials in the grid has been minimized and that the use of the computer program has permitted the location of this conductor to be optimally determined (that is, the conductor has been added only where required).

Normally, mesh potentials are greater along the outside perimeter of a grid, especially at the corner. These potentials can be controlled by decreasing the distance between grid elements, but if the same spacing is used throughout the grid the interior areas of the grid will, in general, be overprotected. Again, the designer can determine the exact spacing

needed for each area of the grid through the use of the ground mat analysis computer program.

12.5 Input Data Requirements. Like all other system studies, a grounding mat analysis study has specific data requirements. The omission of some critical bit of data during a study can significantly affect the validity of the entire study and its conclusions.

Normally, the first step in any ground mat study is the determination of the soil resistivity. If the soil resistivity varies significantly from location to location, the calculated point potentials should be multiplied by an *adjustment* factor (on the order of 0.8 to 0.9) to compensate for uncertainty about the exact resistivity at any given location. Any extremes in weather conditions that might seriously affect the soil's resistivity should also be examined. This usually applies to droughts that can dry out the soil or unusually severe winters that could freeze the soil below the effective depth of the grid. If the substation surface is covered with a layer of crushed rock, its resistivity must also be determined.

The next logical step is the determination of the maximum permissible mesh potential. Many industries have established a standard value for the maximum mesh potential. Otherwise, the estimated fault duration and resistivity of the material at the surface must be used to calculate the maximum safe mesh potential from Eq 5. If a layer of crushed rock has been applied to the surface, the correct value of soil resistivity should be used to calculate the maximum safe mesh potential for locations without surface treatment.

To determine the "total" fault current, the maximum future value of the available symmetrical ground fault current

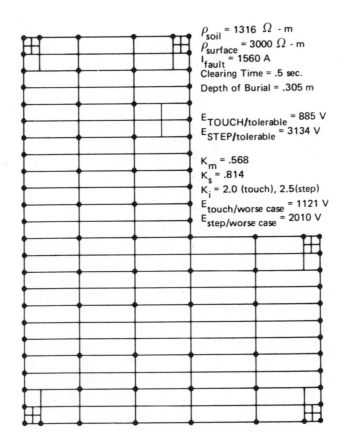

ρ_{soil} = 1316 Ω - m
$\rho_{surface}$ = 3000 Ω - m
I_{fault} = 1560 A
Clearing Time = .5 sec.
Depth of Burial = .305 m

$E_{TOUCH/tolerable}$ = 885 V
$E_{STEP/tolerable}$ = 3134 V

K_m = .568
K_s = .814
K_i = 2.0 (touch), 2.5(step)
$E_{touch/worse\ case}$ = 1121 V
$E_{step/worse\ case}$ = 2010 V

Fig 105
Typical Ground Mat Design Showing
All Pertinent Soil and System Data

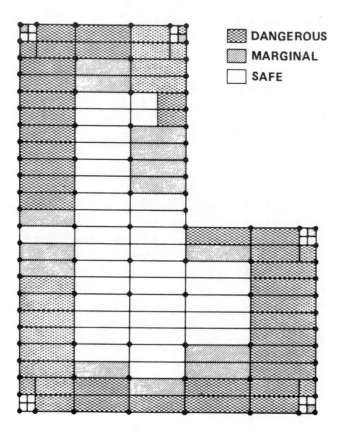

Fig 106
Typical Ground Mat Design Showing
Meshes with Hazardous Potentials
as Identified by Computer Analysis

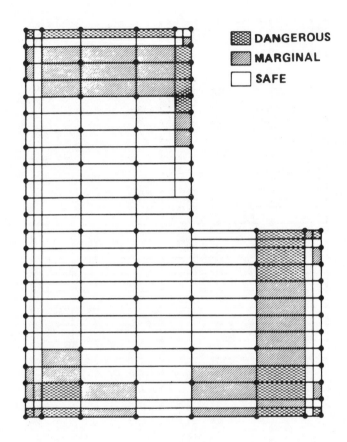

DANGEROUS
MARGINAL
SAFE

Fig 107
Typical Ground Mat Design, First Refinement
Showing Meshes with Hazardous Touch Potentials

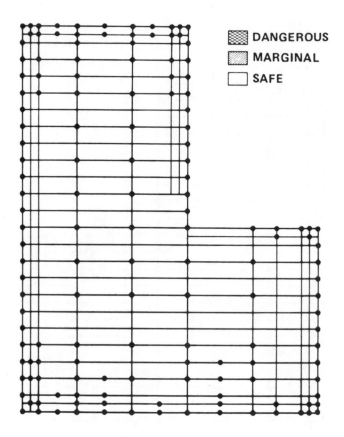

Fig 108
Typical Ground Mat Design, Final Refinement
with no Hazardous Touch Potentials

must be calculated first. Next, the system frequency, X/R ratio, and fault duration should be used to calculate the decrement factor. The decrement factor and the symmetrical ground fault current can then be multiplied by one another to obtain the *total fault current* used by the program. The fault current depends on the grid resistance, which in turn is dependent on grid design, which is constantly being modified. Accordingly, normal practice is to use one of the simplified formulas to find a preliminary value for the grid resistance, then calculate a final value after the grid design has been fixed.

Finally, a tentative grid design must be laid out showing the diameter, depth of burial, and location of each grid conductor. This layout should show any special restrictions on the design such as conductors whose positions have already been fixed by equipment placement, or areas where conductors cannot be located. Any buried electrically conductive object (a pipe, for example) that crosses the grid should be marked on the diagram. The points where the voltage rise will be calculated must be selected. For evaluating touch potentials these normally include the mesh centers, the area underneath the operating switches of all major equipment, and the entrances into the substation. In addition, a point should be considered one meter away from each grid corner along the line bisecting the 270° angle formed by the perimeter grid conductors (see Fig 109). The potential of this point is important for evaluating either step or touch hazards, depending on the location of the perimeter conductors relative to any metallic fencing [6].

In summary, a grounding mat study requires, as a minimum, the following data:

(1) Soil resistivity, both at the surface

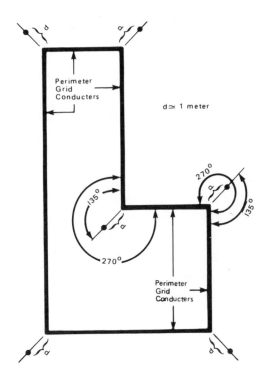

Fig 109
Typical Ground Mat Layout Showing Possible Locations of Critical Step and Touch Potentials Near Grid Corners

and at the level of the grid

(2) Resistivity of any special soil surface dressing material

(3) Estimated duration of a ground fault

(4) System frequency

(5) System X/R ratio

(6) Maximum symmetrical ground fault current, both future and present

(7) Grid layout showing the precise location of every conductor

(8) Coordinates where the potential rise must be calculated

Consideration of all this information will lead to a reliable, accurate, and useful study.

12.6 Typical Computer Output.

12.6 Typical Computer Output. Figure 110 shows a typical output report from a ground mat analysis computer program. In this particular case the output results apply to grid E of Fig 104. Note that both absolute potential and mesh potential are given several different values of fault current at each point of interest in the grid. As described earlier, step potential between any two points (one meter apart) can be determined simply as the difference in the absolute potentials of these points. The output report from such a program can also include the calculated value of grid resistance, the maximum allowable step and touch potentials, and the total length of conductor needed to construct the grid.

12.7 Conclusion. The adaptation of classical analytical techniques and calculating procedures to the digital computer has made grounding mat analysis much more precise, reliable, and useful. While all but eliminating unnecessary grid overdesign, a grounding mat analysis by computer program can detect unsafe conditions that might otherwise go undiscovered until made apparent by serious mishap.

Although grounding mat analysis programs provide an invaluable design tool, they are by no means infallible. If at all possible, a follow-up investigation should be made of each grid after it has been installed. This should include a measurement of grid resistance at the very least, and preferably the measurement of the ac mesh potential at several locations within the grid. If these measured values differ appreciably from the calculated ones, the results of the grid study should be rechecked and supplemental rods or buried conductors provided as required to establish safe conditions [6].

12.8 References

[1] DALZIEL, C. F. Dangerous Electric Currents, *AIEE Transactions*, vol 65, pp 579–585 and 1123–1124.

[2] DALZIEL, C. F. Threshold 60-cycle Fibrillating Currents, *AIEE Transactions*, vol 79, pp 667–673, 1960.

[3] DALZIEL, C. F. and LEE, W. R. Re-evaluation of Lethal Electric Currents, *IEEE Transactions Industry General Applications*, vol IGA-4, pp 467–476, Sept/Oct 1968.

[4] DALZIEL, C. F. A Study of the Hazards of Impulse Currents, *AIEE Transactions*, vol 72, pp 1032–1043, 1953.

[5] FERRIS, L. P., KING, B. G., SPENCE, P. W. and WILLIAMS, H. B. Effect of Electrical Shock on the Heart, *AIEE Transactions*, vol 55, pp 498–515 and 1263, May 1936.

[6] IEEE Std 80-1976, IEEE Guide for Safety in AC Substation Grounding.

[7] DAWALIBI, F. and MUKHEDKAR, D., Optimum Design of Substation Grounding in a Two Layer Earth Structure, *IEEE Transactions Power Applications Systems*, vol PAS-94, No 2, pp 252–272, Mar/Apr 1975.

[8] SUNDE, E. D. *Earth Conduction Effects in Transmission Systems.* New York: VanNostrand, 1949.

[9] RUDENBERG, R. Grounding Principles and Practice I-Fundamental Considerations on Ground Currents, *Electrical Engineering*, vol 64, pp 1–13, Jan 1945.

[10] LAURENT, P. General Fundamentals of Electrical Grounding Techniques, *Bulletin de la Societe Francaise des Electriciens*, vol I, series 7, pp 368–402, July 1951.

[11] NIEMANN, J. Changeover from High-Tension Grounding Installation to

Operation With a Grounded Star Point, *Electrotechnische Zeit*, vol 73, No 10, pp 333–337, May 15, 1952.

[12] SCHWARTZ, S. J. Analytical Expression for Resistance of Grounding System, *AIEE Transactions*, vol 73, pp 1011–1016, 1954.

[13] GROSS, E. T. B., CHITNIS, B. V. and STRATTON, L. J. Grounding Grids for High-Voltage Stations, *AIEE Transactions*, vol 72, pp 799–810, 1953.

[14] GROSS, E. T. B. and WISE, R. B. Grounding Grids for High-Voltage Stations-II, *AIEE Transactions*, vol 74, part III, pp 801–809, 1955.

[15] GROSS, E. T. B. and HOLLITCH, R. F. Grounding Grids for High-Voltage Stations-III, Resistance of Rectangular Grids, *AIEE Transactions*, vol 75, part III, pp 926–935, 1953.

[16] GEDDES, L. A. and BAKER, L. E. Response to Passage of Electric Current Through the Body, *J. Association Advancement of Medical Instruction*, vol 2, pp 13–18, Feb 1971.

[17] DAWALIBI, F. and MUKHEDKAR, D. Multi-Step Analysis of Interconnected Grounding Electrodes, *IEEE Transactions Power Application System*, vol PAS-95, pp 113–119, Jan/Feb 1976.

[18] KOCH, W. Grounding Methods for High-Voltage Stations With Grounded Neutrals, *Elektrotechnische Zeit*, vol 71, No 4, pp 89–91, 1950.

GROUND GRID VOLTAGE STUDY

EXAMPLE PROBLEM
FIGURE 0, GRID E

JANUARY 23, 1974

RESISTIVITY = 1000.0

GRID COORDINATES (MILLIMETERS)		FAULT CURRENT (AMPERES)					
		500.		1000.		4000.	
		VOLTAGE					
		A=ABSOLUTE POTENTIAL B=TOUCH POTENTIAL					
		A	B	A	B	A	B
(30,0,	30,0)	1372,	951,	2743,	1902,	10973,	7610,
(40,0,	30,0)	1307,	1016,	2613,	2032,	10453,	8129,
(40,0,	40,0)	1418,	905,	2836,	1809,	11344,	7238,
(7,5,	67,5)	1984,	339,	3968,	678,	15871,	2711,
(7,5,	82,5)	1975,	348,	3950,	696,	15798,	2784,
(7,5,	97,5)	1902,	381,	3803,	762,	15534,	3048,
(7,5,	112,5)	1851,	472,	3701,	944,	14805,	3777,
(22,5,	67,5)	2015,	307,	4031,	615,	16123,	2459,
(22,5,	82,5)	2023,	300,	4046,	599,	16184,	2398,
(22,5,	97,5)	2001,	322,	4001,	644,	16005,	2577,
(22,5,	112,5)	1936,	387,	3871,	774,	15484,	3098,
(37,5,	67,5)	2019,	304,	4037,	608,	16150,	2432,
(37,5,	82,5)	2032,	291,	4064,	582,	16254,	2326,
(37,5,	97,5)	2016,	307,	4032,	614,	16128,	2454,
(37,5,	112,5)	1963,	360,	3926,	720,	15702,	2880,
(52,5,	67,5)	1941,	342,	3881,	684,	15846,	2736,
(52,5,	82,5)	1975,	348,	3950,	696,	15800,	2782,
(52,5,	97,5)	1960,	363,	3920,	726,	15679,	2903,
(52,5,	112,5)	1931,	392,	3862,	784,	15447,	3136,
GRID POTENTIAL		2323,		4646,		18562,	

NOTE: GRID IS 120X120 MILLIMETERS, ORIGIN (0,0)
IS LOCATED AT LOWER LEFT HAND CORNER.

Fig 110
Sample Output Report—Computer Calculated Ground Grid Potentials

13. Computer Services

13.1 Introduction. The bulk of this book is concerned with the details of power system analysis by the use of computers. Very little, to this point, has been said about computers, programs, or requirements for their use. Many readers are not aware of the wide variety of computational aids available, or factors that are important in choosing the most appropriate aids for their particular needs. This chapter discusses computer systems, services, their use, and the availability of such services. Factors impacting the selection of such things as *in-house* systems, commercial services, time sharing, batch systems, and other decisions which must be made when selecting the most appropriate computing aids are discussed.

The efficiency of these computing tools can be increased considerably if the user has an idea of what to expect and is well prepared before beginning his analysis. This section will strive to acquaint prospective computer users with the problems of data preparation, job submittal, turn-around time, output interpretation, and with the sources of aid when problems arise.

13.2 Computer Systems. There are basically two types of computer systems available to prospective users

(1) In-house company owned or leased systems

(2) Commerical computing services available in a wide variety of types

13.2.1 In-House Systems. The in-house system is normally used by firms with a considerable amount of data processing requirements. Most of these systems are general purpose installations with resources shared by several company departments (for example, Engineering, Accounting, Project Control, etc). Sometimes a company requires special purpose computer equipment to support its engineering operations, and can buy or lease analog or hybrid computer systems. These systems are used almost exclusively for technical applications. Resources, however, are shared among various technical disciplines.

The decision to select an in-house system must consider many important factors. In-house systems are expensive to buy or lease and to operate. Whether bought (and depreciated) or leased, the costs continue even when not in use. The equipment, which includes not only the computer, but magnetic tape and disk drives for bulk data storage, card reader, printer, and other peripherals, must be located somewhere. This usually means a special room, with special environmental considerations such as temperature and humidity controls, as well as grounding and electro-magnetic interference isolation. Computers are, for the most part, very sensitive to perturbation in their power source and can require an uninterruptible power supply to permit adequate reliability. Input medium must be provided (for example, punched cards, paper tape, terminals, etc). This means additional equipment such as keypunch machines or remote terminals. All equipment must be maintained, and in the case of hardware failures, highly trained computer technicians must be called in. This is usually provided for through a maintenance contract. In most shops, except for very small installations, the equipment is operated by specially trained operators. The software (for example programs) is developed, and maintained by a group of full-time computer programmers. The programs can be developed in-house or obtained from outside sources through lease or purchase. Also, they often are obtained from government agencies, or from universities for the cost of reproduction. In summary, in-house systems require large capital investment in equipment, and continued investment in operation and maintenance. Normally there is the need to maintain a group of specialists to operate the equipment and to maintain the software.

13.2.2 Commercial Computing Services. The main advantage of commercial computing services is that they offer all the capabilities of in-house systems with little or no capital expense. Input and output can be by remote terminal or via mail. These services range from the simple use of a computer on which the user installs his own software, to a total consultant service for which the user only supplies the data and receives the answers. The equipment requirements range from none to mini-computer type intelligent terminals, the most common equipment being teletype or remote batch entry terminals, any of which can be leased or purchased. The programs can be developed by the user or the computing service. Also, programs can be obtained from the sources mentioned in the above paragraph for in-house systems.

In general the cost of using a commercial service increases with the amount of service received. It is important to realize that commercial services must recover all costs incurred on an in-house system, as well as return a profit. Therefore, the cost per computation is higher with the services than with a well run, *fully utilized*, in-house system. The key here is the usage requirements. Those firms with low usage requirements can pay many times the actual computer costs for services and still save money. Some firms have computers which are under-utilized. They can be used for other purposes, and power systems analysis programs can be installed on them.

13.3 Types of Computing Service. In this section computing services refers to either a commercial computing service or to an in-house (or captive service) department.

There are two basic operating modes

for computer programs, batch and on-line. Batch programs are initiated and executed by submitting a job via card deck, or terminal input. Once initiated, a batch program runs to completion without interfacing with the user. On-line, or time sharing programs, are executed from terminals. They may require the user to interact by responding to prompting by the computer through the remote terminal.

Batch and on-line programs are available either on commercial services or on in-house systems. In either case, program initiation is via remote terminal (for example, batch entry card reader, or on-line time sharing terminals). Batch programs are run by the service upon receiving the user data in the mail. On-line programs are more costly in the use of computer resources and in general cost the user more per computation than batch programs. On-line programs do, on the other hand, allow more timely answers, which means reduced turnaround time. The user can quickly scan the results and make changes to the data if necessary for subsequent runs. Turnaround time is defined as the elapsed time between submittal of a problem and receiving the output results.

As mentioned previously, the amount of service varies greatly in the amount of consultation available between computing services. A greater amount of service implies a higher cost. At one extreme, programs obtained from *free* sources might not have enough documentation. In these cases input and output format can be determined from analyzing the program code and the usability of the output format may be questionable. On the other end of the spectrum is the total *hands-off* analysis by a consultant type computer based on the user's specifications and data. In this case, the consultant comes to the user and helps find the necessary data for the analysis. The consultant then performs the computer study and informs the user of the result. This type of study is quite expensive when compared to studies in which the user is more involved. Between these extremes is a wide variety of services which can be piece meal or complete analysis reports. In a piece-meal analysis the results of initial computer runs must be manually transcribed and input to later runs. For example, load flow analysis results are required for short circuit, motor starting, and stability program input data. Complete analysis systems allow results of these initial studies to be stored and reused as input, with only the additional data necessary for the later study required from the user.

Commercial services include user manuals and input data sheets (for batch programs) and in most cases some amount of consultation for the user. This consultation is very valuable in improving the effectiveness of using a service. The availability of consultation varies between services and should be considered when selecting a commercial service. This consultation should be sought early in the analysis to avoid problems and wasted man-hours and computer runs.

13.4 Use of Computing Services. After deciding which type of computing service is most appropriate for the user's present needs, several service representatives should be contacted by mail or phone. The larger national computing services have representatives in most large cities and will gladly send information or a representative on request. A user should study available user manuals and company literature to determine if the computing service in question has clear and complete instructions for data entry. They should also give some explanation on how to interpret the output, or results.

In general, input data is the same as that needed for hand calculations. It may, however, be required in a different form than is immediately available to the user. Some programs require data in percent or per unit on a defined base, while others can be on the specific base of the piece of equipment. The user must consult the user manual and give the input data as required. It should be noted that most problems encountered through a computer program are caused by errors in input data. One should be extremely careful when converting data to the form required by the program. Data should be as complete as possible whether submitted via mailed input data sheets, time sharing terminal, or card deck to a batch card reader. The results will be only as good as the input data. The computer has no intelligence and can only work with the data given, in exactly the manner prescribed by the program. Most computing services have a consultation service to aid the user in defining input data requirements and to help him debug the input data in case of program failure. The user should take full advantage of all consultation available from the computing service.

Once the input has been entered the time required for a response (turnaround time) will be a function of not only the type of program (batch or online) but the time of day. Computing services have busy hours (prime time) which are normally described in their literature. Turn-around is usually better during non-prime time hours.

Output reports from most commercial computing services are complete and well labeled to aid interpretation. User manuals normally have a section explaining output reports. What is lacking in many user manuals is trouble-shooting advice if the results appear to be not as expected. All results of computer programs should be scrutinized for answers which are contrary to what the user's experience would predict. If the cause of the discrepancy is not obvious from the printout or cannot be discerned from the user manual, the computing service's consultant should be contacted. Reports can be mailed to the user or printed directly at his location, depending on the amount of hardware the user has.

13.5 Availability of Computing Services. Computing services are available to anyone who has access to a telephone. Telephone lines are the normal means of communication when remote terminals are used. These lines can be leased or paid for on a usage basis. In many cases additional transmission hardware is required which is usually available from the telephone company.

Depending on the size of the computing service being used, a large library of analysis programs is available. Most power system analysis programs contain short circuit, load flow, stability, and motor starting programs. The complexity and sophistication of these programs or program systems vary widely. The user must be careful to select a service whose programs meet his needs. For example a load flow program which can only handle 50 buses may be of no use to a large utility, while being perfectly suited for a small industrial application. In general, the larger the computing service, the wider the selection of available programs. Large services normally have programs of varying complexity and sophistication to satisfy more customers.

For some users the best decision is to

contract with a general service agency which can perform most of the analysis. These services are available from engineering consultants or some of the large equipment manufacturers. They permit an analysis as complete as the user requests and can give advice or opinions if desired. This type of service is more expensive than performing the analysis in-house, but is desirable if time or manpower constraints dictate so.

Much of the information used in the preparation of this section was obtained from literature published by companies interested in data processing equipment and computing services.

Index